GAO

Report to the Ranking Member,
Committee on Natural Resources,
House of Representatives

May 2012

URANIUM MINING

Opportunities Exist to Improve Oversight of Financial Assurances

GAO

Accountability ★ Integrity ★ Reliability

May 2012

URANIUM MINING

Opportunities Exist to Improve Oversight of Financial Assurances

Why GAO Did This Study

From 2005 through 2007, uranium prices increased from about $20 a pound to over $140 a pound, leading to renewed interest in uranium mining on federal land. This interest has raised concerns about the potential impacts that more uranium operations could have on the environment. GAO was asked to (1) compare key agencies' oversight of uranium exploration and extraction operations on federal land, (2) determine the number and status of uranium operations on federal land, (3) identify the coverage and amounts of financial assurances for reclaiming current uranium operations on federal land, and (4) examine what is known about the number and location of abandoned uranium mine sites on federal land and their potential cleanup costs. GAO reviewed agency reports and regulations, surveyed relevant agency field staff on the status of these operations, and examined federal data on uranium operations, financial assurances, and abandoned uranium mine sites.

What GAO Recommends

GAO recommends, among other things, that federal agencies better coordinate their efforts when establishing financial assurances and develop a consistent definition for abandoned mine sites. The Departments of the Interior, Agriculture, and Energy, along with NRC and the Environmental Protection Agency (EPA), concurred with these recommendations. In addition, Interior and EPA provided technical comments, which GAO incorporated as appropriate.

View GAO-12-544. For more information, contact Anu K.Mittal, (202) 512-3841, mittala@gao.gov

What GAO Found

The Bureau of Land Management (BLM), the Forest Service, and the Department of Energy (DOE) are the key agencies that oversee uranium exploration and extraction on federal land, but GAO identified three areas where their oversight processes differ. First, these agencies have different processes for notification of uranium exploration or extraction activities on federal land. Second, the agencies require operators to have in place financial assurances to cover the full estimated cost of reclaiming a uranium operation, but they differ in who estimates the value of the financial assurance and the frequency of their reviews of the assurances. Third, under existing authorities, DOE can collect royalties or rents for uranium extraction, but BLM and the Forest Service cannot. DOE has collected about $64 million in rents and royalties from its leasing program since the 1940s.

As of January 2012, a total of 221 uranium operations were on federally managed land, but only 7 were actively extracting uranium and all of these were on BLM land. An additional 29 uranium operations were awaiting federal approval. Of the 202 operations on BLM land, the majority were engaged in either reclamation or exploration activities, according to BLM field officials. In addition, 3 uranium operations were on Forest Service land, and 16 operations were on lease tracts that DOE manages, none of which were actively extracting uranium.

As of January 2012, BLM, the Forest Service, and DOE reported having $249.1 million in financial assurances, and these assurances were generally adequate to cover the estimated reclamation costs for uranium operations on federal land. Nearly all of these assurances ($247.6 million) were for authorized uranium operations on BLM-managed land, with the remaining $1.5 million for authorized operations on Forest Service land and for DOE's lease tracts. BLM and the Nuclear Regulatory Commission (NRC), which is responsible for overseeing some aspects of uranium operations on federal land, do not coordinate efforts to establish and review financial assurances for in situ recovery operations, which use a series of wells to extract uranium. Such operations account for a large percentage of the total financial assurances held by the agencies.

Federal agencies do not have reliable data on the number and location of abandoned uranium mine sites on federal land or a definitive cost for their cleanup. There are likely thousands of abandoned uranium mine sites on federal land, but GAO identified significant limitations in agencies' data that make their databases generally unreliable. For example, these databases do not have complete data and do not use a consistent definition of an abandoned mine site. Agencies do not know how many sites will need cleanup, and they do not have information on the total cost to clean up these sites. Based on agencies' experiences with cleanup at some sites, cleanup costs could vary significantly from thousands to hundreds of millions of dollars, depending on site-specific conditions and the amount and type of work required at each site.

Contents

Abbreviations

AMSCM	Abandoned Mine-Site Cleanup Module
BLM	Bureau of Land Management
CERCLA	Comprehensive Environmental Response, Compensation, and Liability Act
DOE	Department of Energy
EIA	Energy Information Administration
EPA	Environmental Protection Agency
ISR	in situ recovery
MRDS	Mineral Resources Data System
NEPA	National Environmental Policy Act
NRC	Nuclear Regulatory Commission
UIC	underground injection control
USGS	U.S. Geological Survey

United States Government Accountability Office
Washington, DC 20548

May 17, 2012

The Honorable Edward J. Markey
Ranking Member
Committee on Natural Resources
House of Representatives

Dear Mr. Markey:

From 2005 through 2007, uranium prices increased from about $20 a pound to over $140 a pound, which led to renewed interest in uranium mining—both exploration and extraction—on federal land in the United States. In early April 2012, prices were about $50 per pound, but thousands of claims have been filed to explore for and potentially extract uranium on federal land. This increase in claims filed—the first step in a potentially lengthy process to explore and extract uranium—has raised concerns about the potential impacts that an increased level of uranium exploration and extraction could have on the environment. For example, during uranium extraction, the waste rock piles that are formed can introduce radionuclides (such as radium) and heavy metals (such as selenium and arsenic) into the environment. Before the mid-1970s, many mines on federal land, including uranium mines, were abandoned without any reclamation, leaving a costly legacy of abandoned mines that pose potential health, safety, and environmental hazards. Some of these hazards include open or concealed mine openings, unstable mine structures, and toxic or radioactive materials. In 2008, we reported that from fiscal years 1998 to 2007, the federal government had spent billions to reclaim abandoned hardrock mines, which include uranium mines.[1]

To mitigate these potential health, safety, and environmental hazards, mining operators are responsible for addressing safety hazards and reclaiming the site after their operations have ceased.[2] Activities that

---

[1]GAO, *Hardrock Mining: Information on Abandoned Mines and Value and Coverage of Financial Assurances on BLM Land*, GAO-08-574T (Washington, D.C.: Mar. 12, 2008).

[2]Under the Federal Land Policy Management Act of 1976, the Bureau of Land Management (BLM) issued regulations, effective in 1981, that required all mining operators to reclaim BLM land disturbed by hardrock mining. In 2001, BLM regulations began requiring all mining operators to provide financial assurances before beginning exploration or mining operations on BLM land. The Forest Service began requiring reclamation and financial assurances in 1974.

GAO-12-544 Uranium Mining

address safety hazards can include installing gates over mine openings. Reclamation activities can include reapplication of topsoil, and reshaping and revegetation of disturbed soil areas; measures to control erosion, landslides, and water runoff; measures to isolate, remove, or control toxic materials; and rehabilitation of fisheries and wildlife habitat.[3] Operators are required to obtain financial assurances to cover estimated reclamation costs, and the federal government can use these assurances to pay for reclamation activities if the operator does not reclaim the site.[4] Our past work has raised concern about the adequacy of financial assurances to cover potential reclamation costs for hardrock mining operations, including uranium, on federal land.[5]

A number of federal agencies are involved in the oversight of uranium mining activities on federal land. The Department of the Interior's Bureau of Land Management (BLM) and the Department of Agriculture's Forest Service regulate mining on public domain lands under the General Mining Act of 1872 and other federal land management laws, including the Federal Land Policy Management Act of 1976. The Department of Energy (DOE) administers a uranium leasing program on land that has been withdrawn from the public domain under the Atomic Energy Act of 1954. In addition, the Nuclear Regulatory Commission (NRC) regulates a newer form of extraction, known as in situ recovery (ISR), as a form of uranium milling. The Environmental Protection Agency (EPA) oversees or participates in the remediation of some abandoned mines and sets

---

[3]See 43 C.F.R. § 3809.420 (2011); 36 C.F.R. § 228.8 (2011).

[4]These financial assurances, also referred to as bonds, include a variety of financial instruments. For example, a surety bond is a third-party guarantee that an operator purchases from a private insurance company approved by the Department of the Treasury. The operator must pay a premium to the surety company to maintain the bond. These premiums can vary depending on various factors, including the amount of the bond and the assets and financial resources of the operator, among other factors.

[5]GAO, *Hardrock Mining: BLM Needs to Revise Its Systems for Assessing the Adequacy of Financial Assurances*, GAO-12-189R (Washington, D.C.: Dec. 12, 2011); *Abandoned Mines: Information on the Number of Hardrock Mines, Cost of Cleanup, and Value of Financial Assurances*, GAO-11-834T (Washington, D.C.: July 14, 2011); *Hardrock Mining: BLM Needs to Better Manage Financial Assurances to Guarantee Coverage of Reclamation Costs*, GAO-05-377 (Washington, D.C.: June 20, 2005).

environmental standards for certain sites.[6] Federal agencies may also work with state agencies in overseeing uranium activities. For example, federal agencies may share responsibilities with states for reviewing financial assurances.

You asked us to provide information on the status of uranium mining on federal land. Our objectives were to (1) compare BLM, the Forest Service, and DOE oversight of uranium exploration and extraction operations on federal land; (2) determine the number and status of uranium operations on federal land; (3) examine the coverage and amounts of financial assurances in place for reclaiming current uranium operations on federal land; and (4) examine what is known about the number and location of abandoned uranium mines on federal land and their potential cleanup costs.

To compare how BLM, the Forest Service, and DOE oversee uranium exploration and extraction operations on federal land, we reviewed these agencies' regulations and associated guidance and spoke with agency officials about their implementation of these regulations.[7] In addition, we reviewed NRC and EPA regulations that are relevant to uranium operations and spoke with officials from those agencies. We also reviewed memorandums of understanding among the agencies that delineate their coordination and cooperation in regulating uranium operations, and we spoke with state mining and environmental quality officials to discuss their coordination with federal agencies. To determine the number and status of uranium operations on federal land, we analyzed data from BLM's LR2000 database, which is used to collect and

---

[6]Remediation refers to the containment or treatment of hazardous substances. Remediation work at a mine site can involve removing contaminated waste rock or soil to an off-site location. In addition, contaminated surface or underground water may need to be remediated using a water treatment facility. As described above, reclamation activities may include measures to isolate, remove, or control toxic materials, work that could also be characterized as remediation. Environmental remediation could also be required after a mine is abandoned if reclamation was not properly completed or if new or unforeseen problems arise after abandonment.

[7]We did not include tribal lands in our review of uranium operations on federal land. Currently, there are no active uranium mining operations on tribal lands; however, there are abandoned uranium mines on these lands that will require extensive remediation in some cases. We have included an example of the anticipated remediation actions needed at one such site later in our report. In addition, EPA, DOE, NRC, the Bureau of Indian Affairs, and the Indian Health Service are implementing a 5-year plan to address the health and environmental impacts of uranium contamination in the Navajo nation.

store information on BLM land and programs, including hardrock mining operations. In addition, we administered a web-based survey to all BLM field staff with responsibilities for uranium operations and asked them to provide the status of these operations. Because the Forest Service and DOE oversee fewer operations, we did not send them our web-based survey, but instead reviewed agency documents and interviewed staff from these agencies to determine the number and status of the operations that they oversee.

To examine the financial assurances in place for uranium operations on federal land, we analyzed data and available reports from BLM, the Forest Service, and DOE. We also interviewed officials from these agencies on the processes in place to review financial assurances. As part of this analysis, we examined whether the financial assurances in place were adequate to cover the estimated costs of reclamation; we did not determine whether the estimated costs of reclamation were sound. To learn about the number and location of abandoned uranium mines on federal land, we reviewed data and interviewed officials from BLM, the Forest Service, EPA, the National Park Service, and DOE, which are all involved in efforts to clean up abandoned uranium mines. To assess the reliability of these data, we reviewed documentation from these agencies on their data and interviewed officials involved in collecting and compiling these data. We determined that these data were not sufficiently reliable. Because these data were the only federal data available, we used them to discuss in general terms the potential number of abandoned mine sites, and we describe the limitations of these data. To describe the potential cleanup costs posed by these mines, we identified a series of key cleanup categories that we and agency officials believe are representative of the types of actions that may be required at an abandoned mine. These cleanup categories include actions to (1) address safety hazards, (2) conduct surface reclamation, and (3) remediate environmental hazards.[8] Cleaning up an abandoned mine may involve work that falls across several of these cleanup categories. To provide a range of potential costs for such cleanup work, we asked federal officials for information on past work done to clean up abandoned uranium mines or, if no past work was available, we asked for detailed estimates. We conducted this performance audit from June 2011 through

---

[8]For the purposes of describing the work conducted on abandoned uranium mines, we are using the term "cleanup" to encompass the variety of activities necessary to address conditions at abandoned mine sites.

May 2012 in accordance with generally accepted government auditing standards. Those standards require that we plan and perform the audit to obtain sufficient, appropriate evidence to provide a reasonable basis for our findings and conclusions based on our audit objectives. We believe that the evidence obtained provides a reasonable basis for our findings and conclusions based on our audit objectives. A more detailed description of our scope and methodology is presented in appendix I.

Background

Uranium is a hardrock mineral, and most U.S. uranium deposits are located in the western half of the United States, specifically in the states of Arizona, Colorado, New Mexico, Texas, Utah, and Wyoming.[9] In the United States, uranium has been primarily used as a fuel for electric power generation and for nuclear weapons. In 2010, U.S. uranium mines extracted 4.2 million pounds of uranium, 2 percent more than in 2009, according to DOE's Energy Information Administration (EIA).[10] However, domestic production of uranium is not sufficient to meet domestic demand, and the United States imports over 90 percent of its uranium from countries such as Australia, Canada, and Russia.

Hardrock mining operations consist of four primary stages—exploration, extraction, mineral processing, and reclamation. Several of these stages can take place simultaneously, depending on the characteristics of the operation. Exploration involves prospecting and other steps, such as drilling, to locate mineral deposits. Extraction generally entails developing the mining infrastructure (power, buildings, and roads) needed for extraction, as well as drilling, blasting, and hauling ore from mining areas to processing areas. During processing, operators crush or grind the ore and apply chemical treatments to extract the minerals of value. The

[9]Under U.S. mining laws, minerals are classified as locatable, leasable, or saleable. The General Mining Act of 1872 17 Stat. 91 (codified at 30 U.S.C. § 22 et. seq.) allows individuals to stake claims for locatable minerals, such as uranium, copper, lead, zinc, magnesium, gold, and silver. For the purposes of this report, we use the term "hardrock minerals" as a synonym for "locatable minerals." The Mineral Leasing Act of 1920, 41 Stat. 437 (codified at 30 U.S.C. § 181) created a leasing system for certain minerals such as coal, gas, oil and other fuels, and chemical minerals, which are known as leasable minerals. In 1955, the Multiple Use Mining Act of 1955, 69 Stat. 367 (codified at 30 U.S.C. § 601) removed common varieties of sand, stone, and gravel from development under the Mining Act, and these minerals are known as saleable minerals.

[10]Production data are for pounds of uranium oxide (U_3O_8) extracted from federal, state, and private land.

material left after the minerals are extracted—waste rock or tailings (a combination of fluid and rock particles)—is then disposed of, often in a nearby pile or tailings pond. As described earlier, reclamation activities can include reshaping and revegetating disturbed areas; measures to control erosion; and measures to isolate, remove, or control toxic materials. While uranium mining operations are similar to other hardrock mining operations in environmental concerns, the wastes produced require additional environmental controls. Of particular concern is the presence of the natural by-products of uranium radioactive decay, most notably radium and the radioactive gas radon, as well as heavy metals, such as arsenic. All of these byproducts can pose a serious risk to human health or the environment, especially if they migrate to surface or ground water, or enter the environment after transforming into dust.

Uranium is extracted using one of three processes—underground mining, open pit mining, or ISR. Open pit and underground mining are generally considered conventional uranium extraction processes. In these processes, uranium ore is removed from the ground and is sent to an off-site processing facility, called a mill, where extracted uranium is concentrated into a product called yellowcake (U_3O_8).[11] The optimum extraction process is determined by the size, grade, depth, and geology of an ore body. Open pit mining is generally used for ore deposits relatively close to the surface, while underground mining is generally used for deeper deposits, as shown in figure 1. Open pit mining generally involves more surface disturbance than underground mining, and the amount of waste rock removed to reach the mineral is greater. Since the early 1960s until recently, most uranium has been extracted by using conventional extraction processes.

---

[11]At the mill, the mined uranium ore is crushed, ground, and then fed to a leaching system that uses resin and chemicals to separate uranium from the ore. The resulting yellow slurry—called yellowcake—is washed, dried, and stored in steel drums. Yellowcake subsequently undergoes a number of processing steps (conversion, enrichment, and fuel fabrication) to become fuel for nuclear power plants.

Figure 1: Open Pit and Underground Uranium Mining

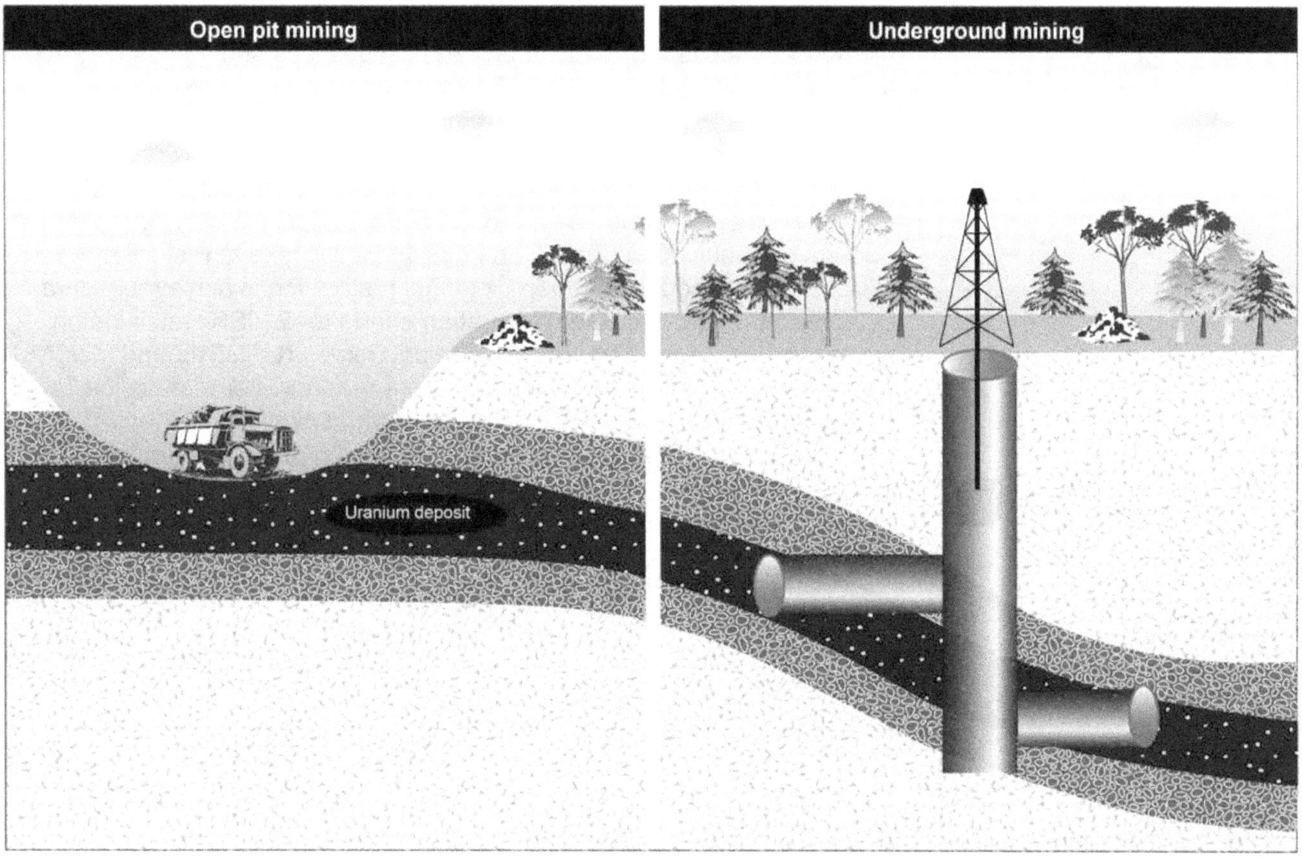

Sources: GAO analysis of information from EPA; Art Explosion (clip art).

Unlike conventional extraction processes, ISR, a mining technique established in the 1970s and anticipated to become more widely used by the industry in the future, aims to extract uranium with less surface disturbance. ISR extracts uranium by injecting oxygenated water and carbon dioxide or sodium bicarbonate hundreds of feet underground to dissolve uranium located in a subsurface ore body contained within a layer of sedimentary rock. Once dissolved, the water and uranium mixture is pumped to the surface, where the uranium is captured on ion exchange resins, which are taken to a central facility to be processed into yellowcake. (See fig. 2.) ISR operations typically involve several wellfields, which are composed of many injection and production wells, and these wellfields can spread over hundreds or thousands of acres, with monitoring wells at periodic intervals above, below, and surrounding the aquifer to monitor for groundwater contamination outside the aquifer. According to industry and government documents, ISR is gaining favor as the approach to extract uranium because it is a more cost-efficient method for recovering uranium ore that causes less surface disturbance and is safer for worker health.[12] The primary risk associated with ISR operations is the potential for contamination of nearby groundwater. When ISR operations cease, the groundwater is restored by removing and stabilizing hazardous metals, such as arsenic and selenium, which may have been disturbed by the operations, and all the wells are plugged. Experts currently do not agree on how long it will take to restore a wellfield after production ceases, or if full restoration is achievable. In a 2009 report on groundwater restoration efforts for 22 ISR wellfields on private land in Texas, the U.S. Geological Survey (USGS) found that it was difficult for these operations to restore groundwater to baseline values for heavy metals, such as uranium and selenium.[13] Specifically, USGS reported that measured levels of uranium and selenium increased following restoration efforts in the majority of the wellfields when compared with baseline values.

---

[12]According to EIA, the amount of uranium that can be produced economically at a market price of $50 a pound using ISR—known as a mineral's reserves—is greater than the amount that can be produced through underground and open pit uranium mining. EIA, *U.S. Uranium Reserve Estimates* (Washington, D.C.: July 2010). Current market prices are close to $50 a pound. At a higher market price of $100 a pound for uranium, the reserves for uranium that can be recovered using underground and open pit uranium mining exceed the reserves for ISR, according to this EIA report.

[13]Susan Hall, USGS, *Groundwater Restoration at Uranium In-Situ Recovery Mines, South Texas Coastal Plain* (Reston, Virginia: 2009).

Figure 2: ISR Extraction Process for Uranium

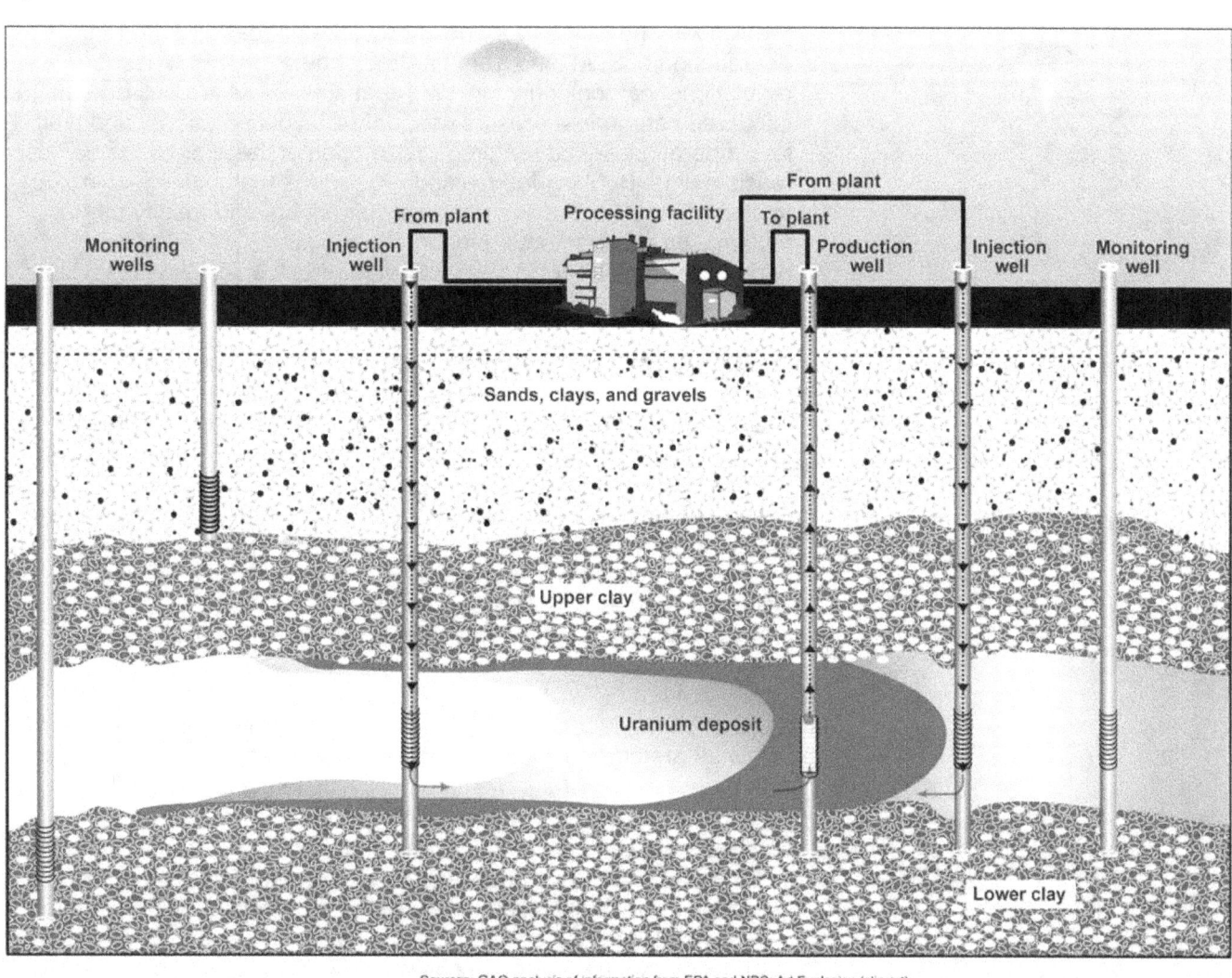

Sources: GAO analysis of information from EPA and NRC; Art Explosion (clip art).

Three federal agencies play key roles in overseeing uranium operations on federal land: BLM, the Forest Service, and DOE. In addition, NRC, EPA, and the states are responsible for some aspects of uranium operations on federal, state, and private land.

- *BLM.* BLM manages more than 260 million acres of public lands located primarily in the western half of the United States. Under the General Mining Act of 1872 (Mining Act), an individual or corporation can establish a claim to any hardrock mineral on public land and may remove all hardrock minerals from the site. Under the Federal Land Policy and Management Act of 1976, BLM has developed and revised regulations and issued policies to prevent unnecessary or undue degradation of BLM land from hardrock operations. BLM issued regulations that took effect in 1981 that classified hardrock operations into three categories—casual use, notice-level operations, and plan-level operations—and required reclamation of the sites at the earliest feasible time. BLM issued revised regulations that took effect in 2001, to strengthen financial assurance requirements and modify the reclamation requirements, among other things. BLM delegates primary responsibility for oversight of hardrock operations to its state and local field offices.

- *The Forest Service.* The Forest Service manages approximately 193 million acres of national forests and grasslands throughout the United States. Forest Service regulations, promulgated under its Organic Act of 1897, among other laws, establish rules and procedures intended to ensure that hardrock mining operations minimize adverse environmental impacts on National Forest System surface resources. Since 1974, the Forest Service has required financial assurances for mining operations on National Forest System land. The Forest Service manages hardrock operations through its headquarters, 9 regions, 155 national forests and grasslands, and more than 600 ranger districts.

- *DOE.* DOE manages a uranium leasing program on 31 lease tracts, of which 29 are currently leased, under the authority of the Atomic Energy Act of 1954 (as amended).[14] These lease tracts cover about 25,000 acres of land located within the Uravan Mineral Belt in southwestern Colorado. These leases generally cover a period of 10 years, and DOE offers these leases through a competitive public bid solicitation, which specifies the lease terms, including the minimum annual royalties to be collected. DOE awards these leases to those operators who offer to pay the highest royalty rate, who become known as lessees. This program began in 1948, when BLM withdrew certain uranium-rich land from the public domain, and reserved them for the use of DOE's predecessor agency, the Atomic Energy Commission, to secure and develop a supply of domestic uranium for the nation's defense needs. DOE manages mining activities, including

---

[14]DOE's regulations are codified in 10 C.F.R. § 760.1 (2012).

exploration and extraction, associated with uranium and vanadium mining on these lands.[15] In 2005, DOE considered an expansion of the program in the face of increased demand for uranium, and initiated an environmental assessment of the program under the National Environmental Policy Act of 1969 (NEPA). DOE subsequently issued a finding that the expansion would have no significant impact on the environment. Environmental groups challenged this finding, and in 2011 a federal court prohibited further work on the leases as well as the issuance of new leases pending completion of a new environmental analysis.[16] DOE is in the process of developing a draft Programmatic Environmental Impact Statement that is expected to be released for public comment in late 2012. According to DOE documents, the lease program has approximately 13.5 million pounds of uranium left to mine.

- *NRC*. NRC is responsible for overseeing uranium milling operations, which produce yellowcake from uranium ore. ISR is considered a uranium milling operation by NRC because it produces yellowcake. NRC reviews ISR license applications, conducts environmental analyses and inspections, reviews decommissioning plans and activities, and oversees site reclamation and groundwater treatment. NRC can relinquish its regulatory authority to a state if the state and NRC determine that the state has a program that is adequate to protect public health and safety. NRC licenses and oversees ISR operations in Nebraska, New Mexico, and Wyoming, while the other states with major uranium deposits—Colorado, Texas, and Utah—license and oversee operations in their states.

- *EPA and the states*. EPA and the states also have a role in overseeing some aspects of uranium operations. Under the Clean Water Act, for example, EPA or the states issue permits to control

---

[15]In the area covered by DOE's leasing program, mined ore contains both uranium and vanadium. This ore is delivered to the processing facility as a combined commodity, and the separate uranium and vanadium minerals are recovered during processing.

[16]On February 27, 2012, the same court ruled that certain reclamation activity, including actions to address dangers to public health and safety and the environment, could continue. *Colorado Environmental Coalition et al. v. Office of Legacy Management et al.* 2012 U.S. Dist. LEXIS 24126 (D. Colo. Feb. 27, 2012).

pollutants that are discharged into the waters of the United States.[17] Under the Safe Drinking Water Act, the Underground Injection Control (UIC) program is designed to protect underground sources of drinking water by prohibiting the injection of fluids beneath the surface without a permit.[18] Specifically, ISR operations require a class III UIC permit for wells because they inject fluids to dissolve and extract uranium. Class III wells must be constructed of appropriate materials to handle the fluid being injected and must be monitored during operations. When injection activities are complete, the injection wells must be plugged. In addition, under the Superfund program, established by the Comprehensive Environmental Response, Compensation, and Liability Act (CERCLA) of 1980, EPA, or, in some instances, other federal agencies if the contamination is on their land, has the authority to compel parties responsible for contaminating sites to clean them up or to clean the sites up itself and seek reimbursement. EPA places some of the most contaminated sites on the National Priority List, and resources from a federal trust fund, the Superfund, are available to pay for long-term cleanup at these sites. In addition, under the Uranium Mill Tailings Radiation Control Act, EPA has established standards for control of radioactive contamination to soil, air, and groundwater at certain uranium processing sites.[19] NRC regulations make EPA's groundwater protection standards generally applicable to uranium milling sites, including ISR operations.

States may play additional roles in regulating uranium operations on federal land. In general, states may have their own requirements governing the review of mining plans, environmental performance standards, reclamation, financial assurances, and inspection. For example, many states with uranium deposits require that an operator provide a financial assurance for the full cost of reclamation for a mining

---

[17]Arizona, Colorado, Utah, and Wyoming have been approved to implement this permit program, known as the National Pollutant Discharge Elimination System program, at the state level. Texas has approval for a partial program.

[18]New Mexico, Texas, Utah, and Wyoming, four states with uranium deposits, have been approved to implement the UIC program at the state level. Colorado implements its UIC program jointly with EPA.

[19]EPA is currently reviewing its existing groundwater standards under 40 C.F.R. pt. 192.

site.[20] Memorandums of understanding among the federal and state agencies aim to encourage coordination between states and federal agencies in overseeing mining operations.

Federal agencies must also comply with NEPA. NEPA requires federal agencies to analyze the likely environmental effects of proposed projects, which may include uranium mines, using an environmental assessment or, if the projects would likely significantly affect the environment, a more detailed environmental impact statement evaluating the proposed project and alternatives. An environmental impact statement results in a record of decision that lays out how anticipated environmental impacts will be mitigated.

Agencies Differ in Their Oversight of Uranium Operations on Federal Land

BLM, the Forest Service, and DOE all oversee uranium exploration and extraction operations on the federal land they manage, but we identified three areas where their processes differ: (1) notification of exploration or extraction operations, (2) oversight of financial assurances, and (3) royalties and rents earned.

BLM, the Forest Service, and DOE Have Different Processes for Notification of Exploration or Extraction

BLM, the Forest Service, and DOE require uranium operators to provide notification of their intent to undertake either uranium exploration or extraction activities on federal land, but their notification processes differ slightly. Under regulations for proposed activities on BLM land, "casual use"—generally defined as activities ordinarily resulting in no or negligible disturbance to the public lands or resources—is allowed without any notice.[21] For operations that are greater than casual use but that will disturb 5 acres or less of land, operators are required to file a notice with the local BLM field office 15 days before commencing operations. Under the regulations, BLM has 15 days to review the notice for completeness. To be complete, a notice must contain specified operator information, a sufficient description and schedule of the activity, a reclamation plan, and a reclamation cost estimate, among other information. Once a financial assurance is in place, the

---

[20]Colorado, New Mexico, Texas, Utah, and Wyoming require financial assurances for the full cost of reclamation, while Arizona does not require financial assurances for the full cost of reclamation.

[21]BLM's regulations for hardrock mining are in 43 C.F.R. subpt. 3809.

operator may begin operations once it hears from BLM that the notice is complete, or if it receives no word from BLM after 15 days. According to BLM guidance, the agency does not approve a notice and therefore is not required to perform an environmental review under NEPA for a notice.

Operations that constitute more than notice-level surface disturbance must submit a plan of operations to the local BLM field office for review and approval, according to BLM regulations. A plan of operations must include, among other information, specific operator information, a description and schedule of operations, a reclamation plan, a monitoring plan, and a reclamation cost estimate. BLM will review the plan within 30 days and then inform the operator that the plan is complete, that more information is required, or that additional steps must be completed. Upon completion of BLM's review of the plan, including analysis under NEPA and public comment, BLM will notify the operator that it approves the plan, approves the plan subject to additional changes or conditions, or that it disapproves or withholds approval of the plan. Since 2001, BLM has been working on a draft handbook to guide its state and local field offices when reviewing notices and plans of operations. In the interim, BLM has issued a series of Instruction Memorandums to its field staff as guidance.

Like BLM, the Forest Service requires operators to provide notification of uranium operations, but the Forest Service differs in the activities it will allow under a notice of intent and plan of operations. Under Forest Service regulations, no notice is required for certain activity, such as collection of mineral specimens using hand tools, but a notice of intent is required for operations that might cause significant disturbance of surface resources, and a plan of operations is required for operations that will likely cause such a disturbance, such as use of mechanized equipment like a backhoe.[22] These standards apply regardless of the acreage involved. Forest Service officials told us that district forest rangers take the lead in reviewing and approving notice- and plan-level operations on Forest Service lands. The Forest Service does not perform environmental analysis under NEPA for projects that are not likely to cause significant disturbance, such as under a notice of intent. A NEPA environmental analysis is initiated only for plan-level operations, because they are more likely to cause significant disturbance.

---

[22]Forest Service regulations governing the surface use of National Forest System land in connection with hardrock mining are in 36 C.F.R. Part 228, subpt. A.

DOE's notification requirements for its lease tracts differ from BLM's and the Forest Service's. DOE officials told us that the majority of its requirements for uranium operations are contained in its bid solicitation and in the terms of the lease, which incorporate relevant sections of DOE regulations. DOE notification requirements for exploration and extraction on its lease tracts are not contained in federal regulations. Instead, our review of two DOE lease documents showed that they contained a section specifying that the operator submit an exploration plan before beginning any surface disturbance to explore, test, or prospect for minerals. Furthermore, the leases specify that before developing a mine, a lessee must submit a separate mining plan to DOE for approval. DOE officials told us that because they oversee operations through a lease, they consider their role to be more like that of a landlord than a regulator. Under a DOE-BLM memorandum of understanding executed in April 2010, DOE has sole authority over the selection of lessees and the negotiation, issuance, management, and termination of leases. However, BLM has jurisdictional authority over all other surface and subsurface uses of the lease tracts and will review and provide comments on lessee plans as they relate to compliance with BLM regulations. According to DOE, it assesses specific tracts through the use of an environmental checklist; however, a more detailed environmental assessment may also take place. DOE reviews mining plans for consistency with its 2007 programmatic environmental assessment and existing environmental regulations.[23] Table 1 describes some of the differences in notification requirements among BLM, the Forest Service, and DOE.

Table 1: Summary of Notification Requirements for Uranium Operations across Three Agencies

Agency	Filing requirement for a notice-level operation	Filing requirement for a plan of operations
BLM	Exploration-related surface disturbance of 5 acres or less	Exploration that disturbs more than 5 acres or any extraction-related operations
Forest Service	Operations that might cause significant disturbance of surface resources	Operations that are likely to cause significant disturbance of surface resources
DOE	Any exploration activity in keeping with terms of lease	Any extraction activity in keeping with terms of lease

Source: GAO analysis of information from BLM, the Forest Service, and DOE.

---

[23]DOE's previous environmental assessment was conducted on the uranium leasing program in 1995.

GAO-12-544 Uranium Mining

BLM, the Forest Service, and DOE Differ in Their Oversight of Financial Assurances

BLM, the Forest Service, and DOE require operators to have financial assurances in place to cover the full estimated cost of reclaiming areas disturbed by operations; however, the agencies differ in who is responsible for initial calculation of these assurances, how frequently they conduct their review, how the review is documented, and how soon reclamation must begin after operations cease. (See table 2 for a summary of financial assurance requirements for the three agencies.) The full estimated cost to reclaim a site is typically defined as the sum sufficient for a third-party contractor to perform all necessary work, including measures to save topsoil for later reuse, control erosion, recontour the area disturbed, and revegetate or reseed the disturbed land. The estimate may also include agency administrative costs.

Table 2: Summary of Financial Assurance Requirements for Uranium Operations across Three Agencies

Agency	Coverage required	Party responsible for initial calculation	Frequency of review	Documentation of review	When reclamation must begin following end of operations
BLM	Full cost of reclamation	Operator	24 months for a notice; 36 months for plan of operations	Documented in LR2000 and summarized annually in Bond Review Report	Promptly for notices; earliest feasible time for plans of operations
Forest Service	Full cost of reclamation	Forest Service	Annually	Recorded in case file, but no agencywide summary of review	Within 1 year, or longer with Forest Service approval
DOE	Full cost of reclamation	DOE	Periodically, or whenever lessee proposes a change in operations	Recorded in case file, but no agencywide summary of review	Promptly and must be completed within 180 days or date agreed to by DOE and lessee

Source: GAO analysis of information from BLM, the Forest Service, and DOE.

BLM regulations require operators to reclaim land disturbed by uranium operations. To ensure that this work is performed, since 2001, BLM has required the operator to provide a financial assurance. Operators must develop an estimate of the amount of financial assurance needed, which BLM reviews and adjusts as necessary. BLM does not have a minimum sum for a financial assurance. BLM uses its Bond Review Report to determine if the estimated costs of reclamation are adequate for ongoing operations, to take action to increase or decrease the financial assurance accordingly, and to certify that financial assurances are adequate to cover estimated reclamation costs. The Bond Review Report aggregates data from BLM's LR2000 database and includes data on the amount of financial assurances and when they were last reviewed. A BLM instruction memorandum directs local field offices to review financial assurances for adequacy every 2 years for notices and every 3 years for plans of operations.[24] In addition, by December 1 of each year, state BLM offices must review the Bond Review Report to determine if reclamation cost estimates for notices and plans of operations within their states are adequate and were reviewed within appropriate time frames. If the Bond Review Report indicates that a financial assurance is not adequate to cover estimated reclamation costs at a site or has not been reviewed within the appropriate time frame, then the state director must develop a corrective action plan to address the deficiencies. Following the end of operations at a site or when a notice expires, BLM regulations require reclamation of a notice to begin promptly, and reclamation of a plan of operations to begin at the earliest feasible time. Because BLM does not have an official definition for these time frames, BLM officials told us that local field offices have flexibility in determining whether operators are in compliance. Before a financial assurance is released back to the operator, the state agency responsible for mine permitting and the BLM local field office will inspect the site to verify that reclamation is complete. In some cases, reclamation can take several years, and a financial assurance may be reduced periodically before being released fully. Because many operations may involve a mix of federal, state, county, and private lands, BLM regulations provide the option of joint bonding with the state.[25] In these cases, the state holds the financial assurance, but it is also redeemable by BLM.

The Forest Service also directs operators to provide a financial assurance for the full cost of reclamation.[26] However, in contrast to BLM, the Forest Service relies on its technical staff at the district, forest, or regional level, not the operator, to calculate the estimated reclamation costs. It uses formal agency guidance issued in 2004 to calculate the estimated reclamation costs and proposes the amount of the financial assurance to cover those costs to the operator. The Forest Service does not have a

---

[24]BLM may review the reclamation cost estimate more frequently if there is cause to believe the reclamation cost estimate is insufficient. A financial assurance for an operation may need to be reviewed annually when it covers an operation that will grow over time according to the timeline submitted in the plan of operations, a practice known as phased bonding.

[25]BLM does not have an agreement covering joint bonding with Arizona.

[26]Forest Service guidance directs its staff to obtain financial assurances to cover the estimated reclamation costs for mining operations on National Forest System lands.

required minimum for financial assurances on its lands. According to Forest Service guidance, an operator's financial assurances should be reviewed annually for adequacy, but a Forest Service official told us that agency staff do not prepare an annual report documenting these reviews. Forest Service regulations require that site reclamation begin upon exhaustion of the mineral deposit, at the earliest practicable time during operations, or within 1 year of the conclusion of operations, unless a longer time is allowed by the Forest Service. Forest Service and state officials will inspect a site to ensure that reclamation is complete before releasing the financial assurance. A financial assurance may also be released in increments as reclamation progresses. In most cases, the Forest Service holds the financial assurances for mining operations on its land, although a Forest Service official told us that the financial assurance could be jointly held with the state for larger operations.

DOE also directs its personnel to ensure that the financial assurance provided by an operator is adequate to cover the estimated cost of reclamation. Sample lease agreements that we reviewed set a minimum financial assurance amount and state that DOE personnel will take into account estimated reclamation costs in setting the financial assurance. Similar to the Forest Service, DOE generally calculates this as the estimated amount for a third-party contractor to perform the reclamation work. The current minimum sum for DOE financial assurances is $5,000, according to DOE officials. Generally, DOE will perform a financial assurance assessment whenever the lessee puts forth new plans for a mining operation. The financial assurance review is filed in the case file as part of the approval package. Upon expiration of the lease, or early relinquishment or cancellation of the lease, current DOE lease terms require lessees to return the site to a condition satisfactory to DOE within 180 days, or a term otherwise agreed to by DOE and the lessee. DOE guidance states that DOE will release the financial assurance once the lessee's reclamation effort is deemed acceptable. Financial assurances are usually held by DOE, except in cases where disturbance to a DOE lease tract is minimal as part of a larger project undertaken on private or state lands.

Unlike BLM and the Forest Service, DOE Earns Royalties and Rents from Uranium Operations

Under existing statutory authorities, BLM and the Forest Service cannot collect rents for the use of federal land or charge royalties on hardrock minerals, including uranium, extracted from that land.[27] BLM does charge claimants an initial $34 location fee, a $15 processing fee, and an annual $140 maintenance fee per claim, and also collects these fees for claims on Forest Service land. In contrast, under the Atomic Energy Act, DOE may collect royalties and rents for uranium extraction operations on its lease tracts. DOE establishes the royalties and terms of payment with the lessee in the lease; typically potential lessees will offer to pay higher production royalties for lease tracts known to contain higher grades of uranium.[28]

DOE has collected approximately $64 million in royalties since the beginning of the lease program in the 1940s. Specifically:

- From the first round of leasing, 1949 through 1962, the program generated $5.9 million in royalties to the federal government from 1.2 million pounds of uranium and 6.8 million pounds of vanadium.

- From the second round of leasing, 1974 through 1994, the program generated $53 million in royalties for the federal government from production of approximately 6.5 million pounds of uranium and 33.4 million pounds of vanadium.

- From the third round of production, 2003 through 2005, the program generated $4.77 million in royalties for the federal government from production of approximately 390,000 pounds of uranium and 1.4 million pounds of vanadium.

---

[27]Unlike BLM and Forest Service, many states provide for the collection of royalty payments. For example, Arizona, Colorado, New Mexico, Utah, and Wyoming charge a royalty for uranium extraction. In the current Congress, the proposed Uranium Resources Stewardship Act (HR1452, 112th Cong. (2011)) would require a royalty charge of at least 12.5 percent on uranium extracted from federal land and rental charges for the land being mined. The money collected would then be used to clean up abandoned uranium mines and mill sites.

[28]The royalty paid differs by lease tract. Leases for tracts held before 2008 require payment of a bid royalty and a base royalty. The bid royalty is a competitive bid made by operators to acquire the lease. The base royalty is set by DOE based on ore production on the lease. Leases rebid on in 2008 require payment of a bid royalty only. The bid royalty is considered the "production royalty" for these lease tracts.

In addition, current DOE leases require lessees to pay an annual rent. According to the program's annual status report, five companies collectively paid an annual rent of $387,040 in fiscal year 2010. Each lessee pays an amount according to the size and value of its lease tract. In lieu of paying this rent, DOE also allows lessees to perform reclamation work on previously abandoned mine sites. In fiscal year 2010, three companies negotiated with DOE to perform reclamation work in lieu of paying rent valued at a total of $101,860.

Over 200 Uranium Operations Are on Federal Land, but Few Are Actively Extracting Uranium

As of January 2012, a total of 221 uranium operations were on federally managed land, but only 7 of these operations were actively extracting uranium and these were all on BLM land.[29] An additional 29 uranium operations were awaiting federal approval. Most of the operations—202—were on BLM land; another 3 were on Forest Service land, and the remaining 16 were on DOE lease tracts.

Uranium Operations on BLM Land Are Generally Engaged in Exploration or Reclamation

Of the 221 uranium operations on federal land, 202, or 91 percent, were on land managed by BLM, according to our analysis of agency data. Of these 202 operations, BLM's LR2000 database identified 144 as authorized, which means BLM has acknowledged an operator's notice or has approved its plan of operations and has approved a financial assurance. These 144 operations included 111 notices and 33 plans of operations, covering about 13,400 acres, and were primarily located in Arizona, Colorado, Utah, and Wyoming. The remaining 58 operations on BLM land were expired notices—that is, operations have ceased except for reclamation and the financial assurance is held until BLM determines that reclamation is complete. According to our analysis of LR2000 data, we also identified 28 uranium operations (11 notices and 17 plans of operations) that were awaiting BLM's authorization. Collectively, these

[29]This count does not necessarily represent individual mine sites because multiple plans of operations may cover a single mine, among other reasons. In addition, the data in this section reflect site status as of January 2012, and the number of uranium operations can fluctuate over time.

pending operations could involve disturbing up to 24,300 acres of BLM-managed land.[30]

We surveyed BLM staff in 25 field offices across eight states for additional information on the status of the uranium operations on BLM-managed land. As shown in table 3, we asked them to provide information on how many operations were in each of eight possible status categories. (For a more detailed description of the status categories that we used in our survey, please see app. I.) Specifically, on the basis of our survey responses, we determined the following:[31]

- Of the 144 authorized operations, 7 operations are actively extracting uranium—3 mines in Utah, 3 in Wyoming, and 1 in Arizona. In addition, 60 operations are engaged in exploration, 51 operations are engaged in reclamation, and 22 are on standby—that is, they are not actively exploring or extracting uranium.[32]

- Of the 58 expired operations, 40 are engaged in reclamation, and BLM staff did not know the status for 12 operations, in part because several of these operations had last been inspected in 2002. Most of the remaining 6 are either in standby or closed status.

- Of the 28 operations identified in LR2000 as pending, field staff reported a status for 12 operations that is inconsistent with BLMs definition of "pending." For example, staff reported 2 pending operations in exploration status, 4 pending operations in reclamation status, 3 pending operations in standby status, and 3 that were closed. Seventeen operations listed as pending in LR2000 were reported by field staff to be in a status that is consistent with the definition of pending, specifically exploration permitting or extraction permitting.

---

[30]This information on acreage reflects the total amount of the authorized area that can be disturbed. However, actual disturbance can often be much smaller, according to BLM officials.

[31]Staff were allowed to select multiple statuses for an operation on our survey. As a result, the sum of responses will exceed the number of operations.

[32]On our survey, we used the terms "mine permitting" and "production." For the purposes of using consistent terms in this report, we are substituting the terms "extraction permitting" and "extraction."

Table 3: Results of GAO's Survey of BLM Field Offices on Status of Uranium Operations

Type of operation	Exploration permitting	Exploration	Extraction permitting[a]	Extraction[a]	Standby	Reclamation	Closed	Other	Don't know
Authorized operations									
Authorized notices	1	55	0	0	7	45	5	2	0
Authorized plans of operations	1	5	2	7	15	6	0	1	0
Subtotal-authorized	2	60	2	7	22	51	5	3	0
Expired operations									
Expired notices	0	0	0	0	1	40	2	3	12
Pending operations									
Pending notices	2	1	0	0	1	4	2	2	0
Pending plans of operations	2	1	13	0	2	0	1	1	0
Subtotal-pending	4	2[b]	13	0	3[b]	4[b]	3[b]	3	0
Total	6	62	15	7	26	95	10	9	12

Source: GAO analysis of BLM field office responses.

Notes: Because an operation could have more than one status, field offices were allowed to select multiple status categories on our survey. As a result, the sum of the responses will exceed the total number of operations. Of the 230 operations, 9 were described by field staff using multiple statuses.

[a]On our survey, we used the terms "mine permitting" and "production." For the purposes of using consistent terms in this report, we are substituting the terms "extraction permitting" and "extraction."

[b]The status reported for these pending operations is inconsistent with BLM's definition of a pending operation.

In addition, our review of documents for 110 of these operations confirmed that some of the reported status levels in LR2000 were inaccurate. For example, we found one notice that was denied in March 2007 that was still listed as pending in LR2000 as of January 2012. In another instance, a notice was authorized in October 2011 but was still listed in LR2000 as pending. There were other instances where the documentation that staff provided to us, such as inspection reports, had not been entered into LR2000. BLM guidance requires that field staff update LR2000 within 5 working days of a change in the status of the operation. Such delays in entering information affect the ability of LR2000 to serve as an effective management tool to track operations. According to the standards for internal control in the federal government, agencies are to promptly record transactions and events to maintain their relevance to management in controlling operations and making decisions.[33]

Of the 7 operations actively extracting uranium on BLM-managed land, 4 are underground mines and 3 are ISR operations. See table 4 for more information on these operations. BLM officials told us the agency did not have data on how much uranium these operations were extracting because it is not authorized to collect this information on uranium or other hardrock minerals.

Table 4: Summary of Operations That Are Extracting Uranium on BLM Land

Operation name	Operator	State	Type of mine
Arizona 1	Denison	Arizona	Underground
Daneros	Utah Energy	Utah	Underground
Pandora[a]	Denison	Utah	Underground
La Sal[a]	Denison	Utah	Underground
Highland[b]	Cameco	Wyoming	ISR
Smith Ranch[b]	Cameco	Wyoming	ISR
Willow Creek	Uranium One	Wyoming	ISR

Source: GAO analysis of BLM data, survey responses, and relevant BLM and company documents.

[a]Both the La Sal and Pandora mines are part of the La Sal Mine complex. We list them separately because they each have separate plans of operations with BLM. The plan of operations for the La Sal mine also includes the Beaver Shaft and Snowball mines. The Pandora mine includes some surface disturbance on Forest Service land resulting from the installation of a few vent holes for the mine; according to Forest Service officials, BLM is the primary federal agency involved in regulating this mine.

[b]The Smith Ranch and Highland operations are adjacent to each other and share a uranium processing facility. We list them separately because they have separate plans of operations with BLM.

Three Uranium Operations Are on Forest Service Land

We identified three uranium operations on land managed by the Forest Service in the Manti La Sal National Forest in Utah. Two of these operations involve uranium exploration, while the third involves the installation of vent holes for the Pandora underground mine, whose entrance is located on BLM-managed land. Collectively, these operations have been authorized to disturb up to 7 acres of land. However, the

---

[33]GAO, *Standards for Internal Control in the Federal Government*, GAO/AIMD-00-21.3.1 (Washington, D.C.: November 1999).

Forest Service is currently reviewing a plan to authorize the Canyon Mine in the Kaibab National Forest in Arizona. This mine's plan of operations was initially approved in the mid-1980s and the Forest Service is determining whether additional, more current environmental analysis must be undertaken to authorize this operation.

All 9 Mines on DOE's Lease Tracts Are on Standby

As part of is uranium leasing program, DOE oversees 31 lease tracts, which are in a variety of statuses.

- Eight tracts have a total of 9 uranium mines on them, all of which are on standby—that is, they are not actively extracting uranium.[34] These lease tracts cover about 6,900 acres, but the operations have disturbed only about 260 acres of land.

- Seven lease tracts have approved exploration plans, but no exploration work is ongoing.

- DOE has not approved any exploration or extraction plans for 14 lease tracts.

- The remaining 2 lease tracts have not been leased out.

According to DOE officials, no extraction activity has taken place on its lease tracts since 2006 for two reasons.[35] First, DOE officials reported that there has been limited incentive to explore or extract uranium on their lease tracts because there are no uranium processing mills in Colorado near the lease tracts.[36] Second, in October 2011, a federal district court ordered that no additional surface disturbance could take place on any DOE lease tracts until DOE completes an appropriate environmental analysis pursuant to

---

[34]One of these lease tracts has 2 mines on it. Of the 9 total mines, 1 is an open pit mine and the other 8 are underground mines.

[35]DOE officials also reported that no exploration activity has taken place on the lease tracts since 2010.

[36]The capacity to process uranium in mills is currently limited, with only one operating uranium mill in the United States, in Blanding, Utah. In Colorado, a uranium mill known as the Piñon Ridge mill is currently in the process of obtaining the necessary permits before it can begin construction.

NEPA.[37] DOE officials told us that a programmatic environmental impact statement is due to be released for public comment in late 2012.

Agency Data Indicate That Financial Assurances Adequately Cover Nearly All Operations, but BLM and NRC Do Not Coordinate in Establishing Some Assurances

As of January 2012, BLM, the Forest Service, and DOE reported $249.1 million in financial assurances, and these assurances appear to be generally adequate to cover the estimated reclamation costs for uranium operations on federal land, according to our analysis of agency data.[38] Agency data indicate that nearly all of these assurances ($247.6 million of the $249.1 million) are for operations that are at least partially on BLM-managed land.[39] Although almost all of these financial assurances were adequate to cover the estimated cost of reclamation, we identified some issues in how BLM oversees these assurances. We also found the value of financial assurances for two ISR operations had increased significantly, but that BLM and NRC did not coordinate their efforts to establish and review financial assurances for these operations. The remaining $1.5 million in financial assurances is for authorized operations on land managed by the Forest Service and for DOE lease tracts. According to our analysis of agency data, these financial assurances are adequate to cover the current estimated cost of reclamation for the operations that the two agencies oversee.

---

[37]In February 2012, the court modified the injunction to allow certain surface-disturbing activities, including those that are absolutely necessary to conduct the environmental analysis.

[38]The data in this section reflect the financial assurances in place as of January 2012, and the value of financial assurances for uranium operations can fluctuate over time.

[39]For operations that involve a combination of BLM and state or private land, BLM's Bond Review Report generally reports the financial assurance for the entire operation, not just the portion on BLM-managed land, and that is what we are reporting. However, BLM does not report information on financial assurances for the portions of mining operations in Arizona that are not on BLM-managed land because the agency does not have a joint bonding agreement with Arizona.

BLM Had Financial Assurances to Cover Reclamation Costs for Nearly All Operations, but Some Issues Exist Regarding Agency Oversight

As of January 2012, BLM had financial assurances of about $245.5 million for 144 authorized uranium operations, according to our review of BLM's Bond Review Report, and the financial assurances were adequate for all but 2 of the operations. Specifically, we found 1 operation where BLM field staff reported that the assurance in place was likely inadequate to reclaim an acid pit lake that had formed at an older, inactive open pit uranium mine in Wyoming. The operation has in place a financial assurance in the amount of $126,000, but the operator is in the process of developing a new reclamation estimate for BLM to review. In addition, we found 1 operation for which the financial assurance for a plan of operations in Utah was $16,000 less than the estimated reclamation costs.[40] In general, we found that most of the financial assurances for operations on BLM land are for less than $100,000.

During our review of BLM's data, we identified two issues related to BLM's Bond Review Report for overseeing financial assurance of uranium operations. First, we found inaccuracies in the information included in the report. Specifically, the Bond Review Report indicated that reviews of the financial assurances for 5 notice-level operations had not taken place in over 36 months, which is a year past the frequency that BLM guidance requires. According to BLM officials, these 5 operations had been reviewed within the correct time frames, but staff had entered an incorrect action code into LR2000. We also found other instances during the course of our review where BLM staff had entered incorrect action codes into this system. LR2000 accepts hundreds of action codes, yet the agency does not have comprehensive guidance on all the action codes that can be used in LR2000.

Second, the Bond Review Report does not include financial assurances that are in place for expired operations. According to our review of agency data, there are 58 expired uranium operations on BLM land. One reason BLM officials offered for why the Bond Review Report does not include information on expired operations was because the financial assurances for these operations are smaller. However, the information we reviewed shows that 43 expired uranium operations had about $2 million in financial assurances and that some of these expired operations had assurances that were well above $100,000. In addition, we found the

---

[40]According to BLM officials, the agency has contacted the new owner of this operation about the need to increase the financial assurance amount.

remaining 15 expired operations did not have any financial assurances in place. According to BLM officials, because these 15 operations were established prior to BLM's 2001 regulations that required financial assurances for all mining operations, it is reasonable that these operations do not have financial assurances. Nonetheless, these 15 operations do need to be reclaimed and, according to BLM staff, these operations may not be receiving the required oversight, which is evidenced by the fact that several of these operations were last inspected about a decade ago. The fact that these 15 operations have not been reclaimed or inspected in almost a decade suggests that oversight of expired operations could be improved.

BLM and NRC Do Not Coordinate when Establishing and Reviewing Assurances for ISR Operations

We found that two ISR operations—the Smith Ranch and Highland operations in Wyoming—account for $213 million in financial assurances, or 86 percent of the total financial assurances held for uranium operations on land managed by BLM. According to BLM officials, a portion of the financial assurances for these two operations also covers activities on land that is not managed by BLM, such as state or private land.[41] The required financial assurances for ISR operations on the Smith Ranch and Highland operations have increased from June 2011 through December 2011—from about $80 million to about $120 million for the Smith Ranch, and from about $80 million to about $93 million for Highland, although the size and disturbance of the operations at these two sites has not significantly changed. According to BLM, NRC, and Wyoming state officials, this increase is due to a variety of factors, including new estimates of the additional work necessary to restore the groundwater at these sites. For example, the estimated number of cycles during which this groundwater is extracted and treated before being reinjected—known as a pore volume—has been increased from six to nine. The cost to restore groundwater at these sites has also increased because the operator had previously removed equipment necessary to restore the groundwater so the equipment could be used in other operating wellfields, and this equipment must now be either returned to these sites or replaced with other groundwater restoration equipment, according to NRC officials. In March 2008, the state of Wyoming issued a notice of violation to the operator for Smith Ranch and Highland that stated that the operator was

---

[41]According to BLM officials, the financial assurances held for uranium operations are not broken out by the entity that manages the surface.

not adhering to the schedule for restoring groundwater and that its estimate of the number of pore volumes and resources needed to restore the groundwater were too low. As a result, the state concluded that the total financial assurances in place at the time for the Smith Ranch and Highland operations—$38.4 million—should be increased immediately to $80 million to protect the public and that a more realistic estimate of the cost to reclaim the sites would be close to a total of $150 million.[42] According to Wyoming state officials we spoke with, this notice of violation was part of the process of requiring greater financial assurances for the Smith Ranch and Highland operations that has resulted in these operations now having a combined $212.7 million in financial assurances.

In examining the efforts to increase financial assurances for these two sites, we found that BLM and NRC did not coordinate their efforts with each other. According to Wyoming state officials, BLM field office staff generally provide comments and concurrence on the proposed financial assurances that operators submit annually. In contrast, NRC generally conducts its own independent review of the financial assurances it believes should be in place. In 2009, NRC and BLM enacted a memorandum of understanding intended to improve interagency cooperation in environmental assessments; facilitate the sharing of special expertise and information; and coordinate the preparation of studies, reports, and documents. However, this memorandum does not cover interagency coordination of the review of financial assurances.

Even though the financial assurances for the Smith Ranch and the Highland operations have increased significantly, the lack of federal coordination when establishing these financial assurances raises concerns about the adequacy of these financial assurances and the financial assurances associated with any future ISR operations that may be authorized. (For more information on active and pending ISR operations, see app. II.) According to our review, it appears that both BLM and NRC have expertise in different areas of the work needed to reclaim an ISR operation, and better coordination among these agencies would help ensure that all necessary factors have been considered. Specifically, BLM primarily has expertise in estimating the cost of reclaiming surface disturbances at a mining site, and NRC primarily has expertise in

---

[42]Wyoming Department of Environmental Quality, "In Situ Uranium Permits 603 and 633, Notice of Violation, Docket No. 4231-08" (Cheyenne, Wyoming, 2008).

estimating the cost of restoring groundwater contaminated by radioactive material. NRC officials reported that some of this expertise was developed through overseeing reclamation activities at uranium processing mills where groundwater must be restored, buildings demolished, and monitoring wells plugged. However, NRC officials acknowledged that the scale of disturbance at an ISR site is much greater than at a mill, because of the thousands of wells that must be plugged and the surrounding surface reclaimed. In addition, restoring the underground water at these mining sites is a complex process because it must be restored to the background concentration, a maximum concentration that incorporates standards set by EPA, or alternate concentration limits as approved by NRC.[43] According to Wyoming state officials we spoke with, enhanced coordination between the federal agencies and also with the state could help to leverage each agency's particular expertise in reviewing financial assurances for ISR sites. These state officials told us that this coordination is even more important because ISR operators have had little experience with restoring groundwater at ISR wellfields to date in Wyoming. Specifically, at the Smith Ranch and Highland ISR sites, the state and NRC have approved groundwater restoration efforts at only 1 of the 19 wellfields according to Wyoming state and NRC officials.

The Forest Service and DOE Have Adequate Financial Assurances to Cover Reclamation Costs for Uranium Activity

The Forest Service and DOE have financial assurances for uranium operations that are adequate to cover the current estimated cost of reclamation for the sites they oversee, according to our analysis of agency data. Specifically, the Forest Service reported having about $42,000 in financial assurances for the three operations on its land, one of which consists of installing vent holes for a mine on adjacent BLM land, and the other two were for operations currently conducting exploration. The Forest Service handbook requires that all active financial assurances be reviewed annually, and our review found that all had been reviewed within appropriate time frames.

DOE reported about $1.5 million in financial assurances for its 29 tracts that have been leased out, with about $1.2 million of this total for a single lease tract with an inactive open pit uranium mine. Our review of DOE

---

[43]Alternate concentration limits can be set if groundwater cannot be restored to background levels, and these limits are based on site-specific conditions at a location. See 10 C.F.R. pt. 40 app. A (2012).

data indicates that these assurances were adequate as of the last time they had been reviewed—from 1996 through 2005 for 9 lease tracts and in 2008 or later for the remaining 22 tracts.[44] DOE officials told us they had not reviewed some of these financial assurances more recently because there has been little new activity on the lease tracts in recent years. DOE officials told us that they generally review financial assurances when a lessee makes a change to an exploration or mining plan on a lease tract.

Federal Agencies Do Not Have Reliable Data on the Number and Location of Abandoned Uranium Mines or Their Associated Cleanup Costs

Federal agencies do not have reliable data on the number and location of abandoned uranium mine sites on federal lands and the potential cleanup costs associated with these sites, according to our review of agencies' databases and discussions with agency staff. We found that agency databases generally lack complete data and a common definition of an abandoned mine site, and contain information that has not been verified through field inspections. In addition, federal agencies do not have estimates of the potential total cleanup cost for abandoned uranium mine sites on the land they manage. According to agency officials, the cost to clean up these sites varies according to site-specific conditions, including the amount and type of work required at each site, and the total number of sites needing cleanup.

Federal Data on Abandoned Uranium Mines Are Unreliable

There are likely thousands of abandoned uranium mine sites on federal land where either exploration or extraction may have taken place, but the available federal data on these sites are generally unreliable. In particular, we found the following limitations with these data.[45]

Agencies' databases are incomplete. Three agency databases only partially track the commodity extracted, and one of them omitted sites with incorrect geographic coordinates. For example, according to BLM's database, there are an estimated 1,189 abandoned uranium mine sites

---

[44]According to DOE, the financial assurances for the 2 lease tracts that were not leased out were last reviewed in June 2008.

[45]DOE also maintains information on abandoned uranium mine sites in a centralized database that also tracks other information related to its uranium leasing program. However, we did not include this database in this analysis, since, according to DOE officials, DOE cleaned up all of its 190 abandoned uranium mine sites from 1996 to 2011 on its lease tracts.

on BLM-managed land. However, these data are based primarily on information from three states (Colorado, Utah, and Wyoming) because the BLM state offices in these states require their local field offices to enter the commodity that had been previously extracted from these abandoned mines.[46] Similarly, in the National Park Service's abandoned mine database, the commodity field is optional for agency staff to enter.[47] On the other hand, EPA's database, which estimates that there are 8,124 abandoned uranium mine sites on federal land, does not include some sites because they do not have specific geographic coordinates, according to agency officials. In addition, some of the databases have not been updated in years and do not track the extent to which extraction took place at each site, which would help indicate the type of cleanup work that might be required. For example, the Forest Service database lists an estimated 1,097 abandoned uranium mine sites; however, the status of many of these sites has not been updated since they were first entered in the database in the 1980s. In addition, the Forest Service and EPA databases do not track which abandoned mine sites have already been cleaned up. As a result, it is not possible to determine from the agency data how many sites remain to be cleaned up.

Agencies do not have a consistent definition of an abandoned mine site. We found agencies do not share a consistent definition of an abandoned mine site, and even within an agency the definition may not be consistently applied by various field offices or staff. These inconsistencies pose a problem when trying to combine multiple databases or to compare data across multiple agencies. For example, because of a lack of a consistent site definition, EPA officials told us that the agency faced a challenge in trying to combine data from multiple sources in order to provide more

---

[46]According to a BLM official, the board in charge of managing the BLM database on abandoned mines has recently decided to eliminate the commodity field from the database.

[47]New hardrock mining claims may not be located on land managed by the National Park Service, and none of the legacy claims currently in the system are for uranium. Therefore, it is highly unlikely that there will be any active uranium operations on National Park Service land. The agency is, however, involved in overseeing efforts to clean up abandoned uranium mines on its land.

GAO-12-544 Uranium Mining

complete information on abandoned uranium mine sites.[48] In addition, even within a single agency, staff may use different definitions of an abandoned mine site when entering data into a database. For example, a BLM official told us that field staff may enter each abandoned mine feature, such as a waste rock pile or a mine opening, as a separate site, instead of grouping these features into one entry. According to a 2007 EPA report on its efforts to develop a database on abandoned uranium mine sites, the lack of a consistent definition leads to problems with determining how many sites exist, since even a single agency's database may contain mines meeting a variety of definitions.[49] In March 2008, we highlighted the lack of a consistent definition for abandoned hardrock mine sites and the way in which this inconsistency contributes to a wide variation in estimates of the number of abandoned mines.[50] At that time, we developed a consistent definition of an abandoned hardrock mine site, and used it to develop a more robust estimate of abandoned mines by applying it across multiple databases. According to EPA officials we interviewed, federal agencies involved with abandoned mines have used a regular interagency forum, called the Federal Mining Dialogue, to discuss the issue of a lack of a common definition of a mine site but have not yet reached agreement on how to address this issue.[51]

Agency databases contain sites that have not been verified through field inspections. According to agency officials, field inspection is the best way to determine an abandoned mine's location and features, such as

---

[48]In 2006, EPA combined data from 19 different databases into one single database. This database primarily includes data from state agencies and BLM's state offices in Arizona, Colorado, New Mexico, Utah, and Wyoming; data from the USGS; as well as some databases with limited number of records from a few states outside these areas, such as California, Montana, South Dakota, and Texas.

[49]EPA, *Technical Report on Technologically Enhanced Naturally Occurring Radioactive Materials from Uranium Mining,* Volume 2 (Washington, D.C.: August 2007).

[50]GAO, *Hardrock Mining: Information on Abandoned Mines and Value and Coverage of Financial Assurances on BLM Land,* GAO-08-574T (Washington, D.C.: Mar. 12, 2008). In this report, we defined an abandoned hardrock mine site as all associated facilities, structures, improvements, and disturbances at a distinct location associated with activities to support a past operation of minerals locatable under the general mining laws.

[51]The Federal Mining Dialogue, established in 1995, is a forum for discussing and coordinating abandoned mine-related issues among federal agencies. EPA serves as the lead agency. Regular participating agencies include BLM, EPA, the Forest Service, National Park Service, and USGS. Other agencies, such as the Department of Justice or the U.S. Army Corps of Engineers, participate when issues of interest arise.

posing physical safety and environmental hazards, to discover new abandoned mine sites, and to figure out what cleanup may be required at an abandoned mine site. However, field inspections also require more resources because agency staff must try to cover large areas of land, sometimes in risky or inaccessible conditions, such as mountainous or rocky areas. Currently, the National Park Service and BLM are in the process of verifying the condition of abandoned mine sites on their land. According to National Park Service officials, the agency received $3.3 million over 3 years to verify how many abandoned mine sites, including uranium mines, it has on the land it manages, and to verify cleanup needs at these sites, a process the agency hopes to complete by September 30, 2012. On the basis of preliminary results from this field inspection, National Park Service officials told us that of the 46 abandoned uranium mine sites on their land, 25 remain to be cleaned up. Since 2009, some inventory efforts of abandoned mines on BLM land have been under way in Arizona, New Mexico, and Wyoming, but not all BLM offices in these states require their staff to track the commodity that was extracted at abandoned mine sites.[52] Table 5 and appendix III provide more specific information on the limitations of each agency's database on abandoned uranium mines.

---

[52]According to BLM officials, BLM was directed to stop any inventory efforts from 1999 to 2009 to focus on cleaning up the already identified abandoned mines because of funding limitations.

Table 5: Limitations with Four Federal Agencies' Databases on Abandoned Uranium Mines

Agency	Database name[a]	Limitations with these databases				
		Partially tracks the commodity extracted	Does not track the extent to which extraction took place at a site	Does not track which sites have been cleaned up	Used an inconsistent definition of a site	Some sites in the database have not been verified through field inspection
BLM	Abandoned Mine/Site Cleanup Module	X	X		X	X
The Forest Service	Forest Service Abandoned Mineral Lands Database	X	X	X	X	X
National Park Service	Servicewide Abandoned Mineral Lands Database	X				
EPA	Technologically Enhanced Naturally Occurring Radioactive Materials Uranium Location Database		X	X	X	X

Source: GAO analysis of information from BLM, the Forest Service, the National Park Service, and EPA

[a]The BLM, Forest Service, and National Park Service databases refer to abandoned uranium mine sites on the land they manage. The EPA database refers to sites on all federal land.

BLM, EPA, and Forest Service officials told us that their agencies do not have an accurate number of abandoned mine sites and their location because no laws or regulations require the agencies to track abandoned mines and that the agencies do not have sufficient resources to collect this information. Specifically, officials from BLM and EPA explained that any tracking of sites is done voluntarily to help with their mission. In addition, BLM and Forest Service officials told us that they have not had sufficient funds to conduct field inspection verification on all their known abandoned mine sites on the lands they manage and that to do so would be costly, requiring additional financial and staff resources. At current funding levels, according to a May 2011 draft feasibility study, it will take BLM 13 years and $39 million to finish inspecting all known abandoned mine sites on its land, including the ongoing inventory work in Arizona, New Mexico, and Wyoming.[53]

---

[53]BLM, *Draft Feasibility Study for AML Inventory Validation and Physical Safety Closures* (Washington, D.C.: May 2011).

Cleanup Costs for Abandoned Uranium Mines Vary Greatly, Depending on Site-Specific Conditions

In addition to not knowing how many abandoned uranium mines are on federal land, BLM, the Forest Service, EPA, and the National Park Service do not have information on the total cost of cleaning up abandoned uranium mines. Officials noted that cleanup costs are determined not only by the total number of mines that need cleanup, but also by site-specific conditions, including the amount and type of work required at each site. Agency officials explained that each abandoned mine site has distinctive characteristics and requires a unique cleanup plan based on, among other things, its size, accessibility, the need for heavy equipment, and the level of contamination.

Agency officials we spoke with generally agreed that cleanup costs at individual sites could range from several thousand dollars to hundreds of millions of dollars. These officials also agreed that most of the work is likely to fall within one of the following three cleanup categories: addressing safety hazards, conducting surface reclamation, and conducting environmental remediation.[54] However, officials cautioned that sometimes cleanup at a site requires work across two or all of these categories. Figure 3 illustrates some of the activities that can take place in these cleanup categories.

---

[54]As discussed earlier, reclamation activities, broadly speaking, may also include environmental remediation. In this section, we distinguish between "surface reclamation," which includes activities such as recontouring and revegetating the land, and "environmental remediation," which involves the containment and treatment of hazardous substances or other toxic materials.

Figure 3: Examples of Cleanup Activities That Could Take Place at Abandoned Uranium Mine Sites

Address physical safety hazards	Conduct surface reclamation	Conduct environmental remediation	
❶ Place gate over mine opening	❶ Remove wooden structure and other debris	❶ Fill the open pit with uncontaminated waste rocks	❹ Long-term treatment of contaminated pit and underground water
❷ Place warning signs	❷ Recontour the site by placing a dry cover over waste rock pile	❷ Remove contaminated waste rocks off-site	❺ Remove off-site contaminated sludge from water treatment
❸ Install fence around the site	❸ Revegetate the site	❸ Revegetate the site	❻ Monitor long-term open pit and underground water

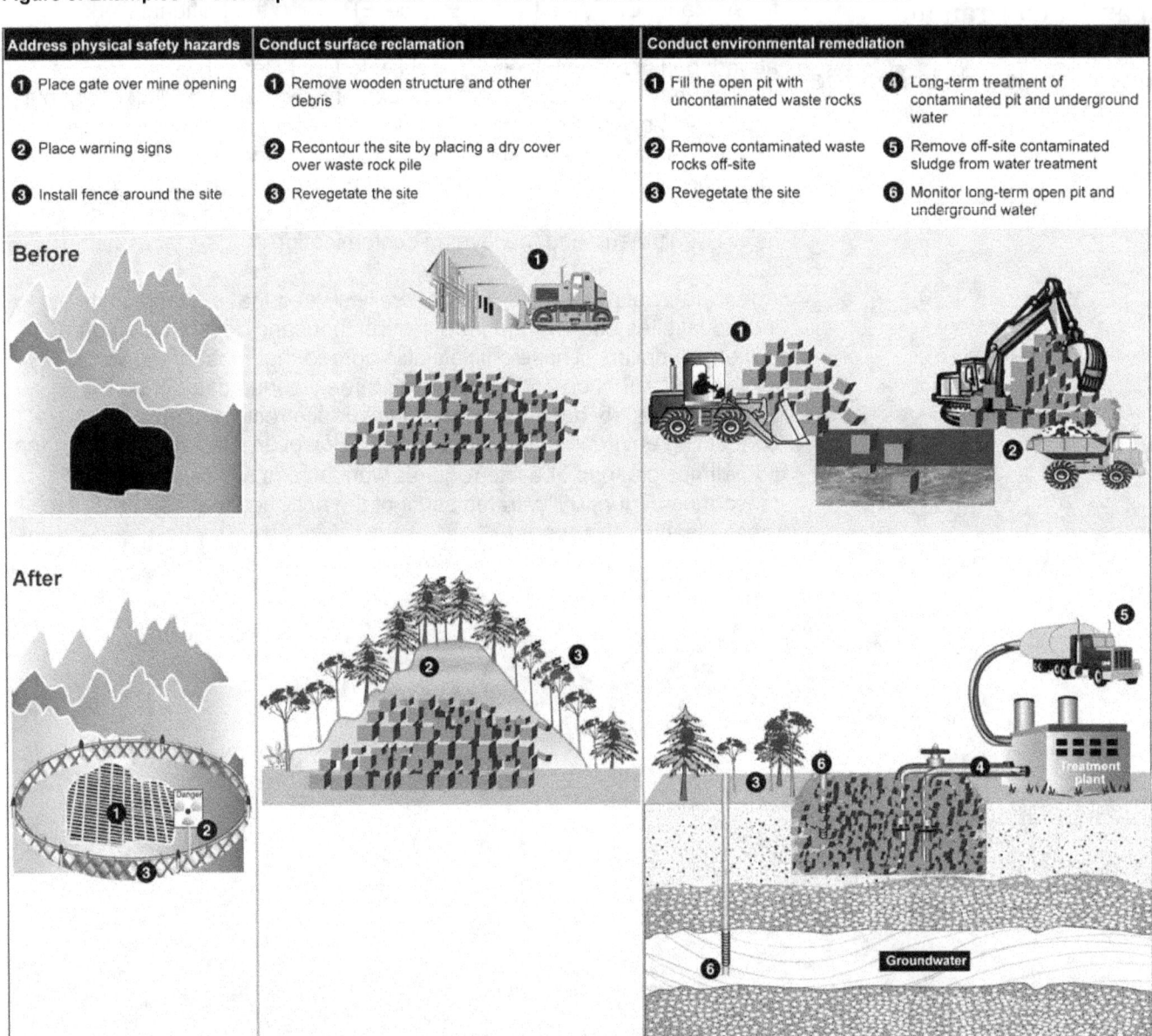

Sources: GAO analysis of information from BLM, DOE, EPA, the Forest Service, and the National Park Service; Art Explosion (clip art).

Note: This figure is illustrative and does not include all possible activities that may take place based on site-specific conditions.

The agencies also provided us with examples of costs that have been incurred at 18 abandoned uranium mine sites. Table 6 provides a range of costs associated with cleanup efforts depending on the type of work conducted at each site. It is important to note that these cost ranges are not exhaustive and that some cleanup costs for other abandoned uranium mine sites could fall outside these cost ranges.

Table 6: Ranges of Costs for Conducting Cleanup Activities at Selected Abandoned Uranium Mine Sites

Primary cleanup work conducted at a site	Number of abandoned uranium sites examined	Range of cost (in 2011 dollars)
Address physical safety	6	$1,800–$33,000
Conduct surface reclamation	6	$2,500–$98,000
Conduct environmental remediation	6	$203,000–$193,000,000[a]

Source: GAO analysis of information received from DOE, the Forest Service, and the National Park Service.

[a]Four of the six examples provided by agencies for this category are based on estimates and not on actual cleanup costs.

Some examples of the factors that can contribute to the variability in the costs for cleanup at abandoned uranium mine sites include the following.

- *Number of safety hazards that need to be addressed:* BLM and National Park Service officials told us that most of the work they have conducted to date on abandoned uranium mines is designed to mitigate safety hazards. Costs for this type of work have ranged from $1,800 to close 2 mine openings in Arches National Park in Utah to $33,000 to backfill 11 mine openings with waste rock at the Canyonlands National Park in Utah.[55] A BLM official cautioned that future costs to address sites with physical safety hazards can be higher because BLM has generally addressed safety hazards that are the least costly to clean up because of limited available funding.[56]

---

[55]The cleanup costs provided in this section are in 2011 dollars.

[56]According to a BLM draft feasibility study, if current funding levels are maintained in the future, it will cost BLM $362.7 million to clean up all of the known abandoned mines, including uranium mines, with physical safety hazards, requiring 77 years.

- *Extent to which surface reclamation needs to be conducted:* The primary purpose of activities under this category is to return the land to as near its previous appearance as possible through recontouring and revegetating disturbed land. According to DOE documents, the costs to reclaim the surface ranged from about $2,500 for closing 2 mine openings, recontouring 70 cubic yards of dirt, and revegetating 1 acre of disturbed land at the Nine Mile Hill Mines on BLM-managed land in Colorado to nearly $98,000 for more extensive reclamation work at the Hawk Mine Complex on lands managed by BLM in Colorado.[57] The work at this site primarily focused on the installation of multiple gates over mine openings, backfilling 500 cubic yards of surface pits with waste materials, recontouring 6,800 cubic yards of waste rock materials from 8 waste rock piles, and revegetating 4 acres of disturbed area.

- *Extent to which environmental remediation must be undertaken:* Most of the activities in this category are designed to mitigate significant environmental hazards. Officials from BLM, the Forest Service, National Park Service, DOE, and EPA told us that few abandoned uranium mine sites have undergone remediation, but cited two instances in which this work has occurred or is ongoing and proved to be costly and the costs varied significantly.[58] For example, according to our review of agency documents, the Pryor Mountain Mine, located on land managed by the Forest Service in Montana, cost about $200,000 to clean up, and involved environmental remediation to remove contaminated soil and waste rock that posed a human health risk. The site, located close to an Indian reservation and near hiking trails and campsites, initially presented levels of radioactive contamination that were up to 369 times higher than normal background levels. At another site—the 320-acre open pit Midnite Mine site in Washington state—costs are estimated to be as high at $193 million by the time remediation is complete, according to EPA documents.[59] Most of this cost is for treating acid rock drainage in two

---

[57]BLM contracted with DOE to conduct this reclamation work on its land.

[58]Both of these examples are from EPA's Superfund program.

[59]This site is located partially on BLM-managed land and tribal lands within the Spokane Indian Reservation. The federal government reached a settlement agreement with mining companies responsible for the site, under which these companies agree to conduct cleanup work and reimburse certain response costs of the federal government. The government agreed to contribute approximately 20 percent of the expected cleanup costs.

large open pits that contain millions of gallons of water and then filling these pits with 33 million tons of waste materials. Some mine sites that require environmental remediation also require long-term— defined as longer than 5 years—maintenance and monitoring, especially if contaminated water requires treatment. For example, one of the largest costs (approximately $32 million) associated with environmental remediation at the Midnite Mine site is for monitoring and treating surface and underground water. EPA estimates that this water will need to be treated in perpetuity.

Additional information on these and other abandoned uranium mine sites is presented in appendix IV.

Conclusions

Having adequate financial assurances to pay for reclamation costs for federal land disturbed by uranium operations is critical to ensuring that the land is returned to its original state if operators fail to complete the reclamation as required. BLM, the Forest Service, DOE, and NRC play key roles in establishing and reviewing these financial assurances for uranium operations on federal land. We found that nearly all of the uranium operations on federal land had adequate financial assurances, according to our analysis of agency data. However, we found some limitations in agencies' oversight of uranium operations' financial assurances, which raise some concerns about these financial assurances. In particular, ISR operations account for a large proportion of financial assurances in place for uranium operations on federal land and have recently been increasing for some operations, yet there is little coordination between BLM and NRC when establishing and reviewing these assurances. This lack of coordination raises concerns about the adequacy of the financial assurances in place for existing ISR operations and for those ISR operations that are awaiting approval. Both BLM and NRC have specific expertise in assessing certain aspects of the reclamation activities that are required at ISR sites, but have no process in place to share this information and leverage their expertise. Without such coordination, the agencies cannot be confident that the assurances they establish for ISR operations will be adequate to cover the costs of reclamation.

BLM relies on its LR2000 database and Bond Review Report to provide information that supports its oversight of financial assurances. However, data entered into LR2000 are sometimes inaccurate and not always updated in a timely manner in keeping with BLM's requirements. Moreover, the Bond Review Report does not examine expired operations, yet we found that some of these operations have large financial assurances in place or have not been inspected in 10 years. Without complete, timely, and accurate information in LR2000 and the Bond Review Report, the usefulness of these management tools to BLM may be diminished and may limit effective oversight of uranium operations.

Finally, identifying the number, location, and cost of cleanup of abandoned mines is a challenging task for federal agencies. However, this process has been made more difficult because the agencies have not been able to reach agreement on a consistent definition for what constitutes an abandoned mine site. Without a consistent definition, data collection efforts are hampered and agency databases cannot be combined to provide a more complete picture of abandoned mines on federal land.

Recommendations for Executive Action

To help better ensure that financial assurances are adequate for uranium mining operations on federal land, we are recommending the following three actions.

- The Secretary of the Interior and the Chairman of the Nuclear Regulatory Commission should enhance their coordination on financial assurances for ISR operations through the development of a memorandum of understanding that defines roles and promotes information sharing.

- The Secretary of the Interior should direct the Director of the Bureau of Land Management to take the following actions to improve oversight of financial assurances:

  - include information on expired mine operations in the annual Bond Review Report process, and

  - develop guidance to ensure accurate and prompt data entry in LR2000.

To enhance data collection efforts on abandoned mines, we recommend that the Secretaries of the Interior and of Agriculture and the Administrator of the Environmental Protection Agency work to develop a consistent definition of abandoned mine sites for use in data-gathering efforts.

Agency Comments

We provided a draft of this report to the Department of Agriculture, the Department of Energy, the Department of the Interior, the Environmental Protection Agency, and the Nuclear Regulatory Commission for review and comment. All of these agencies concurred with our recommendations. In particular, NRC recognized that development of a memorandum of understanding on financial assurance reviews could be beneficial to NRC and BLM, and plans to pursue such an agreement with BLM. NRC noted that development of a memorandum of understanding that adequately addresses both agencies' regulatory oversight may be challenging and stated that the agency may pursue other, less formal methods of coordination with BLM if a memorandum of understanding cannot be developed. In addition, DOE stated that a national database for uranium mining activities would be useful, and the agency agreed there is a need for federal agencies with uranium mines on their land to have common definitions and to use these definitions when gathering information that could be used to determine reclamation needs. Similarly, EPA agreed that a consistent definition of abandoned mine sites would be useful, and will work with other relevant agencies to develop a definition, if possible. Furthermore, EPA commented that our report lacked specificity with regard to our use of the terms "reclamation" and "remediation." We have modified our report to include more specific definitions of each of these terms and clarified what each of these terms means in the context of the report. EPA and the Department of the Interior also provided us with technical comments, which we have incorporated as appropriate. See appendixes V, VI, VII, VIII, and IX for agency comment letters from the Department of Agriculture, DOE, the Department of the Interior, EPA, and NRC, respectively.

As agreed with your office, unless you publicly announce the contents of this report earlier, we plan no further distribution until 30 days from the report date. At that time, we will send copies of this report to the appropriate congressional committees, the Secretary of Agriculture, the Secretary of Energy, the Secretary of the Interior, the Environmental Protection Agency Administrator, the Chairman of the Nuclear Regulatory Commission, and other interested parties. In addition, the report will be available at no charge on the GAO website at http://www.gao.gov.

If you or your staff members have any questions about this report, please contact us at (202) 512-3841 or mittala@gao.gov. Contact points for our Offices of Congressional Relations and Public Affairs may be found on the last page of this report. GAO staff who made major contributions to this report are listed in appendix X.

Sincerely yours,

Anu K. Mittal
Director,
Natural Resources and Environment

Appendix I: Objectives, Scope, and Methodology

Our objectives were to (1) compare Bureau of Land Management (BLM), Forest Service, and Department of Energy (DOE) oversight of uranium exploration and extraction operations on federal land; (2) determine the number and status of uranium operations on federal land; (3) examine the coverage and amounts of financial assurances in place for reclaiming current uranium operations on federal land; and (4) examine what is known about the number and location of abandoned uranium mines on federal land and their potential cleanup costs.

To compare BLM, Forest Service, and DOE oversight of uranium exploration and extraction on federal land, we reviewed federal laws, regulations, and guidance, as well as prior GAO reports and other studies on hardrock mining operations.[1] We also spoke with BLM, Forest Service, and DOE officials in headquarters and field offices, and BLM state offices in Arizona, Colorado, New Mexico, Utah, and Wyoming—five states with large uranium deposits. We also reviewed DOE lease contracts. To understand the interagency relationship among BLM, the Forest Service, and DOE, as well as the these agencies' relationship with the states, we reviewed memorandums of understanding among these parties. We also spoke with state representatives of mining and environmental agencies in Arizona, Colorado, New Mexico, Texas, Utah, and Wyoming to discuss how they coordinate with federal agencies while reviewing uranium operations and their financial assurances. We discussed relevant issues for hardrock operations and financial assurances with representatives from the mining industry, state geological services, and an environmental group. We also examined relevant regulations from EPA and NRC and spoke with officials from these agencies.

To determine the number and status of uranium operations on federal land, we gathered information from BLM, the Forest Service, and DOE. To identify uranium operations on BLM land, we requested that BLM provide an extract from its LR2000 database for operations—both notices and plans of operations—that were in an authorized, expired, or pending

---

[1]We did not include tribal lands in our review of uranium operations on federal land. Currently, there are no active uranium mining operations on tribal lands; however, there are abandoned uranium mines on these lands that will require extensive remediation in some cases. We have included an example of the anticipated remediation actions needed at one such site in our report. In addition, the Environmental Protection Agency (EPA), DOE, the Nuclear Regulatory Commission (NRC), the Bureau of Indian Affairs, and the Indian Health Service are implementing a 5-year plan to address the health and environmental impacts of uranium contamination in the Navajo nation.

status and listed "uranium" or "uranium and other minerals" as the
commodity that was being targeted. To determine the reliability of these
data, we spoke with a BLM information technology official responsible for
administering the system; BLM state and field office staff who enter
information into the system; and BLM managers at the agency's
Washington, D.C., headquarters office who use information from the
system. We also reviewed database documentation, and we determined
the LR2000 data were sufficiently reliable for our purposes. We used
these data to administer a web-based survey to BLM field staff
responsible for overseeing uranium operations in 25 field offices across
eight states—Arizona, Colorado, Nevada, New Mexico, Oregon, South
Dakota, Utah, and Wyoming. We asked these staff to provide the status
of these operations based on the most recent information available using
the following eight status levels and definitions, which we developed in
consultation with BLM staff:

- exploration permitting (e.g., operator is in the process of obtaining
  permits to conduct exploration at the site),

- exploration (e.g., operator is preparing the site for exploration or
  conducting exploration work at the site; concurrent reclamation may
  also be taking place),

- extraction permitting (e.g., operator is in the process of obtaining
  permits to extract uranium at the site),[2]

- extraction (e.g., operator is preparing the site for extraction or actively
  extracting uranium at the site; concurrent reclamation may also be
  taking place),

- standby (e.g. operator is authorized to explore or extract, but is not
  doing so),

- reclamation (e.g., reclamation is taking place at the site following the
  end of exploration or extraction activities),

---

[2]On our survey, we used the terms "mine permitting" and "production." For the purposes
of using consistent terms in this report, we are substituting the terms "extraction
permitting" and "extraction."

- closed (e.g., reclamation is complete and financial assurance has
  been released), and

- other.

As part of this survey, we asked BLM staff to provide copies of the
documentation they consulted when determining the status of the
operation, such as inspection reports or correspondence with operators,
and we used these documents to verify the reported status. For field
offices overseeing a large number of operations, we requested they
provide documents for 10 operations they oversaw, which we selected
randomly. We also asked BLM staff if there had been any uranium
extracted at the operation in the last 5 years. Prior to sending out this
survey, we pretested it with officials from 3 BLM field offices and revised
some of the survey questions based on their input. We received
responses to our survey from all 25 field offices, and we sent follow-up
questions based on their survey responses to clarify certain responses or
to ask for additional information.

Because the Forest Service and DOE oversee fewer uranium operations
than BLM, we did not use our survey to collect information on the status
of these operations; instead, we gathered this information through
interviews with agency officials and agency documents. The Forest
Service compiled information on its uranium operations by contacting
Forest Service officials who were located in National Forests where
uranium operations are located. The Forest Service also provided
documentation on these operations that we used to verify the information
it provided. DOE provided information on its lease tracts that it maintains
as part of its program. We used DOE's annual status report on its lease
tracts to help to verify the reported status levels along with conversations
with DOE officials. For both the Forest Service and DOE, we used
interviews with officials along with relevant documentation to determine
the reliability of these data, and we determined they were sufficiently
reliable for our purposes.

To examine the financial assurances in place for uranium mining on BLM
land, we reviewed information in BLM's Bond Review Report, which
aggregates data on financial assurances from BLM's LR2000 database,
including the required amount of the financial assurance for an operation,
the amount of the financial assurance in place, and when it was last
reviewed. As part of this analysis, we examined whether the financial
assurances in place were adequate to cover the estimated costs of
reclamation; we did not determine whether the estimated costs for

reclamation were sound because that was outside the scope of our
review. Since the Bond Review Report relies on LR2000 data, we used
our data reliability assessment of LR2000 detailed above to help
determine whether the data in the report were reliable. In addition, we
obtained a copy of the specifications that were used to create the Bond
Review Report and examined the report to identify outliers in the data or
incomplete fields and used BLM documents or discussions with BLM staff
to clarify any issues we identified. We determined that BLM's financial
assurance data in its Bond Review Report were sufficiently reliable for the
purposes of our review. Because BLM's Bond Review Report contains
only information on authorized operations, we gathered information on
financial assurances from LR2000 for the expired operations.

To examine the financial assurances in place for uranium operations on
Forest Service land and DOE's lease tracts, we examined data provided
by these agencies. Specifically, we compared the financial assurance
amounts that were required with the amounts that were in place. As we
did for our analysis of BLM's data, we examined whether the financial
assurances in place were adequate to cover the estimated costs of
reclamation; we did not determine whether the estimated costs for
reclamation were sound because that was outside the scope of our
review. To determine the reliability of the data from the Forest Service
and DOE, we interviewed agency staff who gathered these data, and we
used supporting documentation to corroborate the information that was
reported. We determined that these data were sufficiently reliable for our
purposes.

To learn about the number and location of abandoned uranium mines on
federal land, we reviewed data from BLM, the Forest Service, EPA, the
National Park Service, and DOE, which are all involved in efforts to track
and clean up abandoned uranium mines. We received and analyzed data
from databases these agencies maintain on abandoned uranium mines.
We also reviewed pertinent documents that accompanied some of these
databases and other agency documentation, such as studies or reports
that describe the status of abandoned uranium mines on lands managed
or leased by these agencies. We conducted two sets of semistructured
interviews with officials in charge of abandoned mine programs at all of
these agencies—before and after we reviewed the data and
documentation—to gather more information about these databases,
including identifying limitations and determining the reliability of the data
in the databases. We also conducted interviews with officials from the
U.S. Geological Survey, which maintains the data used by the Forest
Service. We also interviewed staff from BLM field offices and state

agencies in the states where most uranium deposits are located to get more information on the number and location of abandoned uranium mines and to hear their perspectives on the federal databases. As a result of our efforts, we determined that these data were not sufficiently reliable to establish a definite number of abandoned uranium mines. However, because these were the only federal data available, we have used them in the report only to discuss in general terms the number of potential abandoned uranium mine sites that may exist on federal lands, and we have described the limitations associated with these data.

To describe the potential cleanup costs posed by abandoned uranium mines, we reviewed relevant literature and conducted semistructured interviews with officials from the federal agencies in charge of abandoned mines.[3] On the basis of this information, we identified three distinct cleanup categories that we and agency officials believe are most representative of the types of actions that take place at abandoned uranium mine sites. In developing these categories, we consulted with officials from all five agencies in charge of cleaning up abandoned uranium mine sites, and they agreed with our approach and our categories. These categories are not mutually exclusive, and cleanup work at a site could fall within multiple categories, especially at larger or more contaminated sites. These cleanup categories included actions taken to

- address safety hazards, which means that most cleanup activities at the site are intended to mitigate safety hazards;

- conduct surface reclamation, which means that most cleanup activities at the site are intended to return the land to its appearance before mining activities took place; and

- conduct environmental remediation, which means that most cleanup activities at the site are intended to deal with removing land and water contamination that poses a threat to the environment and human health. These activities can also include long-term—defined as longer than 5 years—maintenance and monitoring.

---

[3]For the purposes of describing the work conducted on abandoned uranium mines, we are using the term "cleanup" to encompass a variety of activities necessary to address these abandoned mine sites.

We also asked officials from these five agencies to provide us with examples that are illustrative of the range of costs associated with performing such cleanup work. We asked for examples of sites that have already been cleaned up and have definitive costs, or information on sites that have detailed cost estimates. We received 18 examples from the agencies, which are divided equally across the three cleanup categories. Fourteen examples are for past work and contain actual cleanup costs; 4 examples, all in the environmental remediation category, are for work that is still to be completed and are based on estimated costs. For better comparison purposes, we reported these cost numbers in 2011 dollars. For each example, we asked for and received documentation that describes in detail the work performed at each site. For the sites that have not been cleaned up yet, we received pertinent documentation, such as records of decision or consent decrees.

To get a better understanding of uranium mining in general, we conducted site visits to Colorado and Wyoming to examine uranium operations. We visited these states because they have a variety of uranium operations involving several federal agencies. In Colorado, we spoke with BLM, DOE, and state officials involved in overseeing uranium operations. We also spoke with representatives of a uranium company and toured some uranium operations including some underground mines that were on standby on land managed by BLM and a few abandoned mine sites. In addition, we toured two DOE lease tracts and examined reclamation work that had been performed on these tracts. In Wyoming, we met with BLM and state officials involved in overseeing uranium operations and spoke with representatives of some uranium companies. In addition, we toured an in situ recovery operation and examined the various components of this operation.

We conducted this performance audit from June 2011 through May 2012 in accordance with generally accepted government auditing standards. Those standards require that we plan and perform the audit to obtain sufficient, appropriate evidence to provide a reasonable basis for our findings and conclusions based on our audit objectives. We believe that the evidence obtained provides a reasonable basis for our findings and conclusions based on our audit objectives.

Appendix II: Information on In Situ Recovery Operations on BLM Land That Are Extracting Uranium, on Standby, or Awaiting Federal Authorization

This appendix provides information on in situ recovery (ISR) operations on land managed by BLM. Some of these operations are not entirely on federal land, but rather include state and private land. The Forest Service and Department of Energy officials reported that they do not have any ISR operations on land they manage.

Table 7: Information on ISR Operations Located on BLM-Managed Land and Their Associated Financial Assurance

Name	Location	Operator	Financial assurance amount (dollars in millions)	Status
Highland	Wyoming	Cameco	$92.73	Extracting uranium.
Smith Ranch	Wyoming	Cameco	120.04	Extracting uranium.
Willow Creek	Wyoming	Uranium One	16.3	Extracting uranium.
Gas Hills	Wyoming	Cameco	3.47	Waiting for BLM authorization. Authorized by the Nuclear Regulatory Commission (NRC).
Hank and Nichols	Wyoming	Uranerz	6.8	Waiting for BLM authorization. Authorized by NRC.
Lost Creek	Wyoming	UR Energy	2.09	Waiting for BLM authorization. Authorized by NRC.
Reynolds Ranch	Wyoming	Cameco	Smith Ranch financial assurance covers this operation.	Authorized by BLM and NRC but not extracting uranium.
Ross	Wyoming	Strata	No decision on financial assurance yet.	Waiting for BLM and NRC authorization.
Ruth	Wyoming	Cameco	$0.18	BLM has not yet received a plan of operations for this operation. Authorized by NRC.
Dewey Burdock	South Dakota	Powertech	No decision on financial assurance yet.	Waiting for BLM and NRC authorization.

Source: GAO analysis of BLM and NRC information.

Appendix III: Detailed Information on Federal Abandoned Mine Databases

This appendix provides information on federal databases that contain information on abandoned uranium mines, and the limitations that we identified for each database.

Table 8: Information on Federal Abandoned Mine Databases and Their Limitations

Agency	Database name	Number of abandoned uranium mines listed in the database[a]	Mines cleaned up to date listed in the database[a]	Number of abandoned uranium mines in the database that remain to be cleaned up[a]	Limitations with the data
BLM	Abandoned Mine-Site Cleanup Module (AMSCM)[b]	3,038	1,849	1,189	• Entering the mined commodity in this database is optional. Only three BLM state offices (Colorado, Utah, and Wyoming) require BLM staff to enter information on the mined commodity. • The database does not provide information on the extent to which extraction took place at a site. • BLM officials from various field offices who enter information in the database use their own definition of a "site." • Some sites have not been verified through field inspection.
The Forest Service	Forest Service Abandoned Mineral Lands Database, which relies entirely on the U.S. Geological Survey's (USGS) Mineral Resources Data System (MRDS)[c]	1,097	Unknown	Unknown	• MRDS has not been updated since 1995. Also, it does not include information on the major commodity mined for over 22,000 of its records. • The database does not provide updated information on the extent to which extraction took place at a site. • The definition used by the Forest Service for a site is different than the definition used by the USGS for its MRDS database. • Few sites have been verified through field inspection. • MRDS does not identify whether a site has already been cleaned up or not. • MRDS database contains many duplicates.
National Park Service	Servicewide Abandoned Mineral Lands Database	46	21	25	• Entering the commodity field is optional for National Park Service staff.

Agency	Database name	Number of abandoned uranium mines listed in the database[a]	Mines cleaned up to date listed in the database[a]	Number of abandoned uranium mines in the database that remain to be cleaned up[a]	Limitations with the data
EPA	Technologically Enhanced Naturally Occurring Radioactive Materials Uranium Location Database	8,124	Unknown	Unknown	• Some data were not included in the database because they did not have adequate geographic coordinates. • The database does not track the extent to which extraction took place at a site. • The database does not track which mines have already been cleaned up. • The definition of a mine used by the different databases from which the Uranium Location Database was compiled leads to problems with determining how many mine sites exist. • Some data have not been verified through field inspection. • An unknown number of duplicate entries remain in the database.

Source: GAO analysis of information from BLM, the Forest Service, National Park Service, and EPA.

[a]The BLM, Forest Service, and National Park Service databases refer to abandoned uranium mine sites on the lands they manage. The EPA database refers to sites on all federal land.

[b]According to a BLM official, AMSCM is a stand-alone internal database housed at the National Operations Center in Denver, Colorado, which is separate from BLM's larger LR2000 data system.

[c]According to a Forest Service official, the Forest Service is in the process of starting work on its own database on abandoned mines. The new database will also keep track of which sites have been cleaned up to date.

Appendix IV: Examples of Cleanup Activities at Abandoned Uranium Mine Sites

This appendix provides information on cleanup activities at 18 abandoned uranium mine sites. Fourteen sites have been cleaned up and have actual cleanup costs, while 4 examples provided by agencies are based on estimates and not on actual cleanup costs.

Table 9: Examples of Cleanup Activities at Abandoned Uranium Mine Sites

Mine name, location / (federal agency managing the land)	Description of the mine	Summary of cleanup work performed or planned at the site[a]			Total cost (in 2011 dollars)
		Address physical safety hazards	Conduct surface reclamation	Conduct environmental remediation	
Mines where cleanup focused on physical safety					
Salt Valley Wash Mines, Utah (National Park Service)	Two small underground mines, each with one shaft (vertical mine opening) in Arches National Park, dating from the 1940s.	Backfilled two shafts by hand with 24 and 63 cubic yards of adjacent waste material	None	None	$1,818
Loma Mines, Colorado (Bureau of Land Management)	Conventional underground mines located beneath a bluff 2 miles from an interstate highway.	Backfilled two mine adits (horizontal mine opening) with trash and wood debris, and closed them with polyurethane foam	None	None	2,105
Terry Mine, Utah (National Park Service)	Conventional underground mine located in the Capitol Reef National Park, within one-quarter mile of a main road.	Backfilled a shaft with 85 cubic yards and an adit with 50 cubic yards of waste materials Erected two fences of 800 and 960 feet in length to exclude grazing cattle Placed warning signs around the site	Revegetated the site Burned wooden structures	None	10,443
Whirlwind, Utah (National Park Service)	Conventional underground mine located about 400 feet above the elevation of a lake in Glen Canyon National Recreation Area. The cleanup crew and equipment were flown in by helicopter.	Demolished structures Backfilled adits with steel, wood, and other debris Closed a drill hole with polyurethane foam Posted four warning signs	Demolished an 800-square-foot steel ore bin	None	11,448

Mine name, location / (federal agency managing the land)	Description of the mine	Summary of cleanup work performed or planned at the site[a]			Total cost (in 2011 dollars)
		Address physical safety hazards	Conduct surface reclamation	Conduct environmental remediation	
White Rim, Utah (National Park Service)	Four conventional underground mines located in Glen Canyon National Recreation Area. One site is located about 1 mile from the park's Visitors Center and another about 1 mile from a campground.	Installed steel gates over five adits Backfilled by hand one adit using 8 cubic yards of material	None	None	22,499
Lathtrop Canyon 1-8, Utah (National Park Service)	Conventional underground mine located in Canyonlands National Park that was developed in the late 1950s. The crew accessed the site by foot and the equipment was flown in by helicopter.	Closed 11 adits using various methods Installed warning signs throughout the site	None	None	33,021
Mines where cleanup focused on surface reclamation					
Nine Mile Hill Mines, Colorado (Bureau of Land Management)	Conventional underground mines located above a public highway.	Backfilled two small adits with waste rock material using an excavator	Recontoured 70 cubic yards of waste rock materials Revegetated 1 acre of land	None	2,524
Mesa No. 5, Colorado (Bureau of Land Management)	Conventional underground mine that contained a waste rock pile, which affected an adjacent, intermittent stream.	Closed two small-diameter ventilation shafts with polyurethane foam Backfilled a mine opening with 5,200 cubic yards of waste rock materials and trash and debris	Recontoured 10,200 cubic yards of waste rock material Covered the recontoured area with topsoil excavated from a nearby site Revegetated 4 acres of land	None	12,963
Northern Light Mines, Colorado (Bureau of Land Management)	Conventional underground mine.	Placed wood debris from demolished structures in mine openings Installed gates over 3 adits Backfilled 5 additional shafts with waste rock	Recontoured 600 cubic yards of waste rock Covered the recontoured area with topsoil Revegetated 2 acres of land	None	17,121

Mine name, location / (federal agency managing the land)	Description of the mine	Summary of cleanup work performed or planned at the site[a]			Total cost (in 2011 dollars)
		Address physical safety hazards	Conduct surface reclamation	Conduct environmental remediation	
Rainy Day, Utah (National Park Service)	Conventional underground mine located in the Capital Reef National Park 4 miles from the main road. The site was accessible by foot.	Placed debris from demolished structures in mine adits

Backfilled 12 adits with earth from the disturbed slope using a backhoe | Demolished structures

Corrected areas of severe erosion and built drainage controls | None | 30,768 |
| New Verde Mine, Colorado (Bureau of Land Management) | Single underground large mine. This is a historic site where parts of the operation were left in place and preserved. | None | Demolished and disposed of structures offsite

Removed trash and debris offsite

Recontoured 22,000 cubic yards of waste rock materials

Covered the recontoured area with topsoil

Revegetated 6 acres of land | None | 78,524 |
| Hawk Mine Complex, Colorado (Bureau of Land Management) | A complex of underground mines that started operations in 1948, consisting of eight separate and distinct mine sites, six mine access ramps, three adits, and one surface pit. | Placed debris from eight locations in mine openings

Backfilled seven mine openings with waste rock

Installed gates at three mine openings

Closed nine small ventilation shafts and 132 exploration drill holes using polyurethane foam | Backfilled 500 cubic yards of surface pits with waste rock materials

Recontoured 6,800 cubic yards of waste rock materials from eight waste rock piles

Covered the recontoured area with 1,000 cubic yards of topsoil

Revegetated 4 acres of land | None | 97,589 |

Mine name, location / (federal agency managing the land)	Description of the mine	Summary of cleanup work performed or planned at the site[a]			Total cost (in 2011 dollars)
		Address physical safety hazards	Conduct surface reclamation	Conduct environmental remediation	
Mines where cleanup focused on environmental remediation					
Pryor Mountain Mine, Carbon County, Montana (the Forest Service)	Two separate mines developed in the 1950s located in an area where hiking and camping take place. The site is also used as a sacred ground by a nearby Indian tribe. The site was accessible by four-wheel-drive roads, and large equipment was brought in by helicopter.	Installed gates over three adits Backfilled a number of exploration pits Removed 4- to 8- foot highwalls (edge of the mine) Removed one collapsed structure	Revegetated all of the disturbed areas within the site	Removed human health risks related to the site (no details were available of the actual work conducted)	203,238
Workman Creek Uranium Mines, Gila County, Arizona (the Forest Service)	The site, developed in the 1950s, encompasses eight mines on steep hillsides located near campgrounds. Waste rock piles contain elevated levels of carcinogenic and radioactive elements, and there is concern of these materials getting into major water supplies that serve the Phoenix metropolitan area.	Plan to: Install warning signs	Plan to: Backfill the excavated areas with clean soil Recontour some areas to establish stability and prevent water runoff Cut off road access to the site	Plan to: Backfill all 33 mine openings with 30,500 cubic yards of waste rock, some of which is contaminated, and close them with polyurethane foam Remove 500 cubic yards of contaminated waste rock piles from the campgrounds and creek side and place in a repository on site	600,000[b] (estimate)
San Mateo Uranium Mine, New Mexico (the Forest Service)	The site, which operated between 1957 and 1971, is located in a remote location with limited public access. The waste rock pile is toxic and radioactively contaminated, and storm water runoff from this pile flows into a nearby creek.	Plan to: Install an 8-foot-high chain-link fence to enclose approximately 17 acres necessary to exclude wildlife and livestock from destroying the vegetation	Plan to: Recontour and revegetate approximately 35 acres to further reduce windblown transport of any residual contamination	Plan to: Consolidate and cover 180,000 cubic yards of contaminated waste rock pile Conduct ongoing maintenance to repair erosion of the cap material and of the drainage channels after heavy rainfall events	3,095,750[b] (estimate)

Mine name, location / (federal agency managing the land)	Description of the mine	Summary of cleanup work performed or planned at the site[a]			Total cost (in 2011 dollars)
		Address physical safety hazards	Conduct surface reclamation	Conduct environmental remediation	
White King Lucky Lass, Oregon (the Forest Service)	This site consists of two conventional open pit mines for a combined 140 acres of disturbed land. It operated between 1955 and 1965. The site had three large waste rock piles of approximately 1.26 million cubic yards and two large pits that cover approximately 5 and 14 acres, which are full with millions of gallons of water. Carcinogenic and radioactive contaminants were found at the site and a creek runs next to the mine and received discharge from this contaminated pit. This site was added to EPA's National Priority List in 1995.	Installed three-strand barbed wire fencing around wetlands and reclaimed waste stockpiles	Regraded, replaced topsoil, capped with a dry cover system, and revegetated the waste rock piles Restored and revegetated other disturbed areas	Removed contaminated soil from the stockpiles Relocated the flow of the creek into historic channels and constructed three wetland areas to prevent runoff into the creek Installed 10 wells for groundwater monitoring Plan to perform long-term monitoring and regular neutralization of the pit water to prevent any acidic water from reaching the creek	5,939,087
Riley Pass Mine, Harding County, South Dakota (the Forest Service)	This site, which operated in the 1950s and 1960s, involves 12 mine groups and spreads over approximately 1,000 acres. The site consists of numerous open pits, waste rock piles, and five sediment ponds. The area is prone to erosion, and the site poses safety concerns from unstable highwalls. The site also poses health and environmental risks from heavy metals and radiation. Two mines have already been cleaned up and cleanup at a third is ongoing. The U.S. government is currently involved in a bankruptcy proceeding with potentially responsible parties to recover costs for the remaining cleanup.	None	Plan to: Reshape the highwalls Fill or reshape the erosion gullies Stabilize the fragile soils and revegetate the area	Plan to: Establish a series of sediment control measures, such as sediment ponds, to control runoff Excavate and place contaminated material in designed repositories, or cap the contaminated material in place The Forest Service estimates that a long-term maintenance effort will be needed for at least 100 years because of the fragile soil and climate conditions	74,733,520[b] (estimate)

Mine name, location / (federal agency managing the land)	Description of the mine	Summary of cleanup work performed or planned at the site[a]			Total cost (in 2011 dollars)
		Address physical safety hazards	Conduct surface reclamation	Conduct environmental remediation	
Midnite Mine, Stevens County, Washington (Bureau of Land Management)	This is a more than 320-acre open pit mine that operated from 1954 to 1981. Approximately 33 million tons of waste material was dug up from six large pits, two of which have not been backfilled and are full of water. Numerous piles of waste materials are also located throughout the site. High levels of toxic and radioactive chemicals are at the site. Acidic water drains into a nearby creek. Some cleanup work at the site has already been conducted. This site was added to EPA's National Priority List in 2000.	Plan to: Build a fence around the site and boulder barriers around the contaminated waste piles	Plan to: Cover the four pits that were backfilled with a thick dry cover, clean soil, and vegetation Grade and cover waste piles and areas cleaned of waste throughout the site with fresh soil and vegetation Conduct long-term maintenance and monitoring of the dry cover systems and the vegetation to mitigate acid rock drainage	Plan to: Empty out the two pits full with acid water and treat this water at a water treatment plant on site Cover the bottom of these pits with a thick drainage layer where water can collect and install a water removal system along with filling the pits with waste rock and covering the pits with a thick vegetated cover Build a new water treatment plant to treat millions of gallons of acidic groundwater and dispose of sludge Build sediment barriers to prevent sediment migration from the mine drainages into the creek Conduct ongoing maintenance and monitoring of water treatment and remove sludge for at least 140 years	193,000,000[b] (estimate)

Sources: GAO analysis based on information provided by DOE, the Forest Service , and the National Park Service.

[a]It is important to note that we summarized some of the key activities performed at the site; other activities may also have taken place at the site.

[b]This amount is an estimate, since cleanup has not yet been completed or has not started at the site.

Appendix V: Comments from the Department of Agriculture

File Code: 1420/2810
Date: APR 2 6 2012

Ms. Anu K. Mittal
Director, Natural Resources and Environment
U.S. Government Accountability Office
441 G. Street NW
Washington, DC 20548

Dear Ms. Mittal:

Thank you for allowing the Forest Service to review the draft GAO report, GAO-12-544, "Uranium Mining: Opportunities Exist to Improve Oversight of Financial Assurances". The Forest Service has reviewed the draft report and agrees with your recommendations.

Thank you again for the opportunity to review your draft report. If you have any questions, please contact Thelma Strong, Acting Chief Financial Officer, at 202-205-1321 or tstrong@fs.fed.us.

Sincerely,

For Mary Wagner

THOMAS L. TIDWELL
Chief

Appendix VI: Comments from the Department of Energy

Department of Energy
Washington, DC 20585

April 30, 2012

Ms. Anu K. Mittal
Director, Natural Resources and the Environment
Government Accountability Office
441 G Street, NW
Washington, DC 20548

Dear Ms. Mittal:

The U.S. Department of Energy (DOE) appreciates the opportunity to review and comment on the draft Government Accountability Office (GAO) Report, *Uranium Mining — Opportunities Exist to Improve Oversight of Financial Assurances"* dated May 2012 and GAO's incorporation of comments that were presented at our meeting on March 14, 2012. In summary, DOE supports the recommendations in the draft report. Although briefly mentioned, there are some DOE actions related to uranium mining that are pertinent to the report and we believe merit additional discussion. These actions include: historical mine reclamation efforts; the uranium leasing program database, and interagency collaboration to address uranium mining on the Navajo Nation.

Mine Reclamation

Although acknowledged in footnote 43, there is limited discussion of the success that DOE has had in reclaiming 190 abandoned uranium mines on our lease tracts in Colorado. In 1995, the Department began reclaiming the abandoned uranium (legacy) mine sites on our lease tracts within the Uravan Mineral Belt in southwestern Colorado. These legacy mine sites originated with the Atomic Energy Commission's initial mineral leasing program (circa 1948 to 1962).

Early in the reclamation process, DOE recognized that reclamation standards did not exist for legacy mine sites. Accordingly, DOE collaborated with the U.S. Bureau of Land Management (BLM) Colorado State Office and the three BLM field offices located in southwestern Colorado to establish guidelines to be used in the reclamation of legacy mine sites. Those guidelines were issued by the BLM Colorado State Office in November 1995. DOE's reclamation efforts were completed in 2010 with final reclamation of all 190-legacy mine sites located on the lease tracts.

During that same timeframe, BLM officials recognized the Department's experience and expertise in reclaiming legacy mine sites and requested DOE's assistance in reclaiming similar mine sites located on public lands administered by BLM. In April 2000, an interagency agreement was established between the two agencies and during the ensuing 8 years, DOE oversaw the reclamation of 182 legacy mine sites on behalf of BLM.

Printed with soy ink on recycled paper

Database on Uranium Leasing

To support the Department's lease tracts and legacy mine reclamation activities, a geodatabase
was developed to collect, store, and maintain all geospatial data associated with the operation of
the lease tracts. This geodatabase contains location data and attributes for all uranium mine sites
located on DOE's lease tracts, plus similar data for the BLM mines sites where DOE oversaw
reclamation. This database is operating and expanding as new data are collected.

DOE agrees with the audit team that a national database for "uranium mining" would be useful
to combine information compiled by federal and state agencies and Tribal Nations. The process
needs to ensure that the data are compiled into a standardized format that will be reviewed,
revised, and expanded as necessary to enable a comprehensive picture of the nation's uranium
mining activities. DOE further agrees with the GAO audit team, that there is a need for federal
agencies with land that supports uranium mining to have common definitions and to use that
national database to determine reclamation needs.

Federal Agency cooperation on the Navajo Nation

DOE believes that GAO made a small, but important addition to the report, to acknowledge that
the report does not address the issue of abandoned uranium mines on tribal lands. There is a
significant and ongoing interagency effort to address the effects of historical uranium mining on
the Navajo Nation. Five federal agencies, the Environmental Protection Agency (EPA), the
Nuclear Regulatory Commission (NRC), the Indian Health Service, the Bureau of Indian Affairs,
and DOE are in the fifth year of a five-year plan created in 2007. A second five-year plan is
scheduled to commence in 2013.

One important component of that plan is the work associated with the Northeast Church Rock
mine in New Mexico. Both the Navajo Nation and the New Mexico Environment Department
are active participants in this effort. EPA Region IX has proposed, and the NRC and DOE have
agreed, to pursue the co-disposal of mine waste with uranium mill tailings at the Church Rock
mill tailings site in New Mexico. DOE has also agreed to accept a smaller quantity of mine
waste at the Ford Uranium Mill Tailings Radiation Control Act (UMTRCA) Title II site in
Washington State to help facilitate the reclamation of the Midnite Mine on land managed by the
U.S. Forest Service.

The Department believes that we have been successful in inventorying and reclaiming
abandoned uranium mines on our lease tracts. We are committed to working with other agencies
to address the findings and recommendations of the GAO. Should you need additional
information, you may contact Mr. Thomas Pauling at 202-586-1782.

Sincerely,

David W. Geiser
Director
Office of Legacy Management

Appendix VII: Comments from the Department of the Interior

United States Department of the Interior

OFFICE OF THE SECRETARY
Washington, DC 20240

MAY - 3 2012

Ms. Anu K. Mittal
Director, Natural Resources and Environment
U.S. Government Accountability Office
441 G Street, N.W
Washington, D.C. 20548

Dear Ms. Mittal:

Thank you for giving the Department of the Interior the opportunity to review and comment on the Government Accountability Office (GAO) draft report entitled, *Uranium Mining: Opportunities Exist to Improve Oversight of Financial Assurances* (GAO-12-544). The draft GAO report includes three recommendations for the Secretary of the Interior that are intended to strengthen oversight of uranium mining on federal lands.

The Department concurs with the recommendations relevant to its regulatory oversight. Enclosed are some technical comments on the draft report. We hope the comments will assist you in preparing the final report.

If you have any questions, please contact Mitchell Leverette, Chief, Division of Solid Minerals, at (202) 912-7113, or LaVanna Stevenson, BLM Audit Liaison Officer, at (202) 912-7077.

Sincerely,

Rhea Suh
Assistant Secretary
Policy, Management and Budget

Enclosure

Appendix VIII: Comments from the Environmental Protection Agency

UNITED STATES ENVIRONMENTAL PROTECTION AGENCY
WASHINGTON, D.C. 20460

MAY 0 3 2012

OFFICE OF
SOLID WASTE AND
EMERGENCY RESPONSE

Ms. Anu K. Mittal, Director
Natural Resources and Environment
Government Accountability Office
Washington, D.C. 20548

Dear Ms. Mittal:

Thank you for the opportunity to comment on the draft report of the U.S. General Accounting Office (GAO) entitled "Uranium Mining: Opportunities Exist to Improve Oversight of Financial Assurance" (GAO-12-544). The draft report presents one recommendation to the Administrator of the U.S. Environmental Protection Agency (EPA) and other recommendations to the Secretaries of the Interior and Agriculture, and the Chairman of the Nuclear Regulatory Commission. We have responded below to the recommendation presented to EPA. Other specific comments on the draft report are included in the Enclosure.

We appreciate GAO incorporating most of our previous comments on the Statement of Facts which preceded the development of this draft report and for the collegial working relationship and dialogue with EPA staff.

GAO Recommendation

The draft report states: "To enhance data collection efforts on abandoned mines, we recommend that Secretaries of the Interior and of Agriculture and the Administrator of the Environmental Protection Agency work to develop a consistent definition of abandoned mine sites for use in data-gathering efforts."

EPA Response

EPA agrees with GAO's recommendation. EPA suggested this recommendation during the March 28, 2012, Federal Mining Dialogue (FMD) meeting in Washington, D.C. The FMD representatives from EPA, U.S. Departments of Interior (DOI) and Agriculture (USDA) agreed that they will work to develop a consistent definition of abandoned mine sites, if possible. However, the FMD members note that it may be challenging to develop a consistent definition of abandoned mines because of the legal authorities that each agency implements. An example of this challenge is that DOI and USDA may identify, prioritize and address mine safety issues at abandoned mines whereas EPA may not identify or address these types of sites. Another example is that EPA considers mineral processing facilities under our universe of hardrock mining sites but DOI and USDA may not.

General Comments:

In general, we believe that the GAO staff did a reasonable job reviewing the issue and identifying that it is difficult and expensive to address. While the report focuses in part on lack of precise data for the number and size of abandoned mines, it does not discuss the lack of Federal funding being provided to the Agencies to address the issue. Even without precise data, the order of magnitude of the problem described in the existing data bases is sufficient to identify the general scope of the problem.

As further discussed in our specific comments, it is important to provide clearer definitions of the terms reclamation and remediation. This is significant because the report extensively discusses both financial assurances to cover reclamation and remediation at various sites. Reclamation goals are not the same as remediation goals, and because of this the report may be misleading. We think the report should, at the beginning, define these terms and provide the regulatory background for each.

In closing, we believe that there is substantial useful information in this report. Please feel free to contact me or Shahid Mahmud at 703-603-8789 if you have additional questions.

Sincerely,

Mathy Stanislaus
Assistant Administrator

Enclosure

cc: Bob Abbey, DOI
Harris Sherman, USDA
Suzanne Rudzinski, ORCR
Cynthia Giles, OECA
Dan Opalski, EPA Region 10
Barry Breen, OSWER
James Woolford, OSRTI
Susan Bromm, OFA
Bobbie Trent, OSWER
Johnsie Webster, OSWER

2

Appendix IX: Comments from the Nuclear Regulatory Commission

UNITED STATES
NUCLEAR REGULATORY COMMISSION
WASHINGTON, D.C. 20555-0001

April 26, 2012

Mr. Anu K. Mittal, Director
Natural Resources and Environment Division
U.S. Government Accountability Office
441 G Street, NW.
Mail Stop: 2T23
Washington, DC 20548

Dear Mr. Mittal:

Thank you for the opportunity to review and submit comments on the draft U.S. Government Accountability Office (GAO) report GAO-12-544, "Uranium Mining: Opportunities to Improve Oversight of Financial Assurances," which the NRC received on April 9, 2012. The NRC appreciates the time and effort that you and your staff have taken to review this topic.

The GAO reviewed the practices of the different Federal agencies that maintain responsibility for oversight of uranium exploration and extraction operations on Federal land. The Federal agencies involved with oversight of uranium exploration and extraction operations include the Bureau of Land Management (BLM), Forest Service, the U.S. Department of Energy, and the NRC. The GAO determined that, in general, adequate financial assurance is in place to address uranium exploration and extraction operations on Federal land. The GAO recommends that Federal agencies better coordinate financial assurance reviews and develop a common definition for abandoned mine sites. In addition, the GAO provided one specific recommendation for the NRC:

> The Secretary of the Interior and the Chairman of the Nuclear Regulatory Commission should enhance their coordination on financial assurances for ISR operations through the development of a memorandum of understanding that defines roles and promotes information sharing.

The NRC has reviewed the report and specific recommendation to develop a memorandum of understanding (MOU) with BLM for financial assurance reviews. Historically, the NRC and BLM have not coordinated financial assurance reviews. In the more than 30 years of the NRC uranium recovery program, we are unaware of any situations where a problem has occurred resulting from a lack of coordination between NRC and BLM. However, the NRC recognizes that development of an MOU could be beneficial to both agencies. Therefore, the NRC plans to pursue development of an MOU with BLM on financial assurance reviews. This effort would be sequenced with an on-going effort to develop MOU's with States for financial surety reviews.

A. Mittal   2

The NRC believes that development of an MOU that adequately addresses both agencies'
regulatory oversight may be challenging. BLM's focus on uranium exploration and extraction
operations is related to land disturbance, whereas the NRC's focus is on radiological and
chemical contamination and the restoration efforts needed to address those two concerns.
Additionally, substantial land disturbance on Federal land can occur prior to NRC granting a
license. Site exploration, characterization, and pre-construction are allowed under NRC's
regulations before a license is granted. Thus, the timing for having financial assurance in place
can be distinctly different for the two agencies. In some cases, exploration activities may occur
years before issuance of an NRC license.

In the event that an MOU cannot be developed, the NRC may pursue other less formal methods
of coordination with BLM.

The NRC has no additional comments on the report. Should you have any questions about
these comments, please contact Jesse Arildsen of my staff at 301-415-1785.

Sincerely,

R. W. Borchardt
Executive Director
for Operations

Appendix X: GAO Contact and Staff Acknowledgments

GAO Contact	Anu K. Mittal, (202) 512-3841 or mittala@gao.gov
Staff Acknowledgments	In addition to the individual named above, Andrea Brown and Elizabeth Erdmann (Assistant Directors), Antoinette Capaccio, Julia Coulter, Maria Gaona, Scott Heacock, Cristian Ion, Rebecca Shea, Carol Herrnstadt Shulman, and Jena Sinkfield made key contributions to this report.